P9-DCQ-977

The Child in the World/The World in the Child

WITHDRAWN

Critical Cultural Studies of Childhood

Series Editors:

Marianne Bloch (University of Wisconsin–Madison)
Gaile Sloan Cannella (Arizona State University Tempe)
Beth Blue Swadener (Arizona State University Tempe)

This series will focus on reframings of theory, research, policy, and pedagogies in childhood. A critical cultural study of childhood is one that offers a "prism" of possibilities for writing about power and its relationship to the cultural constructions of childhood, family, and education in broad societal, local, and global contexts. Books in the series will open up new spaces for dialogue and reconceptualization based on critical theoretical and methodological framings, including critical pedagogy, advocacy and social justice perspectives, cultural, historical and comparative studies of childhood, post-structural, postcolonial, and/or feminist studies of childhood, family, and education. The intent of the series is to examine the relations between power, language, and what is taken as normal/abnormal, good and natural, to understand the construction of the "other," difference and inclusions/exclusions that are embedded in current notions of childhood, family, educational reforms, policies, and the practices of schooling. Critical Cultural Studies of Childhood will open up dialogue about new possibilities for action and research.

Single authored as well as edited volumes focusing on critical studies of childhood from a variety of disciplinary and theoretical perspectives are included in the series. A particular focus is in a re-imagining as well as critical reflection on policy and practice in early childhood, primary, and elementary education. It is the series intent to open up new spaces for reconceptualizing theories and traditions of research, policies, cultural reasonings and practices at all of these levels, in the United States, as well as comparatively.

WITHDRAWN

The Child in the World/The World in the Child

Education and the Configuration of a Universal, Modern, and Globalized Childhood

Edited by
Marianne N. Bloch, Devorah Kennedy,
Theodora Lightfoot, and Dar Weyenberg

TOURO COLLEGE LIBRARY
Kings Hwy

KH

THE CHILD IN THE WORLD/THE WORLD IN THE CHILD
© Marianne N. Bloch, Devorah Kennedy, Theodora Lightfoot, and Dar Weyenberg, 2006.

All rights reserved. No part of this book may be used or reproduced in any manner whatsoever without written permission except in the case of brief quotations embodied in critical articles or reviews.

First published in 2006 by
PALGRAVE MACMILLAN™
175 Fifth Avenue, New York, N.Y. 10010 and
Houndmills, Basingstoke, Hampshire, England RG21 6XS
Companies and representatives throughout the world.

PALGRAVE MACMILLAN is the global academic imprint of the Palgrave Macmillan division of St. Martin's Press, LLC and of Palgrave Macmillan Ltd. Macmillan® is a registered trademark in the United States, United Kingdom and other countries. Palgrave is a registered trademark in the European Union and other countries.

ISBN-13: 978–1–4039–7497–6 hardback
ISBN-10: 1–4039–7497–7 hardback
ISBN-13: 978–1–4039–7498–3 paperback
ISBN-10: 1–4039–7498–5 paperback

Library of Congress Cataloging-in-Publication Data

The child in the world : the world in the child : education and the configuration of a universal, modern, and globalized childhood / edited by Marianne N. Bloch...[et al.].
     p. cm.—(Critical cultural studies of childhood)
     Includes bibliographical references and index.
     ISBN 1–4039–7497–7 (hc : alk. paper) – ISBN 1–4039–7498–5 (pb : alk. paper)
     1. Comparative education. 2. Education, Elementary—Standards—Cross-cultural studies. 3. Curriculum change—Cross-cultural studies. 4. Education and state—Cross-cultural studies. 5. Education and globalization—Cross-cultural studies. I. Bloch, Marianne N. II. Series.
LB43.C45 2006
372'.9—dc22   2006044795

A catalogue record for this book is available from the British Library.

Design by Newgen Imaging Systems (P) Ltd., Chennai, India.

First edition: October 2006

10 9 8 7 6 5 4 3 2 1

Printed in the United States of America.

11/3/06

Contents

**Part 4   Governing the Modern and Post-Modern
            Citizen and Nation through Universal
            Reforms in Education**

Series Editors' Preface

This book is the first in a series, *Critical Cultural Studies of Childhood*, focusing on reframings of theory, research, policy and pedagogies in childhood. As longtime colleagues in the "reconceptualizing early childhood" movement, we have discussed for many years ways in which critical cultural studies of childhood offer a complex array of possibilities for writing about power and its relationship to the cultural constructions of childhood, family, education, and public policy in broad societal, local, and global contexts. Our intentions for this series include opening up new spaces for dialogue and reconceptualization based on critical theoretical and methodological framings, including critical pedagogy, advocacy and social justice perspectives, cultural, historical and comparative studies of childhood, poststructural, postcolonial (anticolonial), and feminist studies of childhood, family, and education. As such, this series provides a space for scholarship examining the relations between power, language, and what is taken as normal/abnormal, good and natural, understanding the construction of the "other," difference and inclusions/exclusions that are embedded into current and prevailing notions of childhood, family, educational reforms, policies, and the practices of schooling.

The child in the world/The world in the child: education and the configuration of a universal, modern and globalized childhood embodies several of these complex theoretical framings of prevailing discourses and governmentalities related to ongoing reconceptualist and critical cultural studies debates in early childhood as well as elementary education studies. These debates, in particular, focus a critical theoretical attention on the historicity and politics of education as these are reflected in the constructions of childhood, education, schooling, family as universal, and in need of being universal, within an increasingly complex, globalizing world. In particular, this volume relates to increasing calls for standards, the "universal" child, and the "need" for greater accountability and "best practices" in a range of policies affecting children and families. Such policies and advocacy discourses are often based on social prescriptions for populations that comprise the "other." Chapters draw from scholarship in the United States, Japan, Brazil, England, and Taiwan to help frame how the ideas of modernity and a universal child, as well as standards and policies for *all*, travel across borders, and are positioned differently within different localities, and "timespaces." The chapters address early and elementary education issues, with foci on "the child," "the curriculum," and "the world."

Contributing authors draw primarily from Foucauldian and postcolonial perspectives and evoke, with Alvermann (2000) and Deleuze and Guattari (1987), the metaphor of a "rhizomatic" space where it is possible to "question and re-examine the striated space—our commonplace understandings, without being so abstract or so open that our analysis has no connection and no meaning" (Bloch, Kennedy, Lightfoot, & Weyenberg, Introduction, this volume). Using a range of primarily poststructural methodologies, contributors provide textual and discursive analysis of classroom observations, policy documents, popular media, curricula, interview narratives, and other "texts," all of which are linked to power/knowledge relationships and implicit governing patterns.

Representing several disciplines, many of the chapter authors have studied at the University of Wisconsin-Madison, which can be described as one of the birthplaces of the reconceptualist movement in early childhood, and a space that reflects a substantial critical and postmodern curriculum theory legacy. As such, the contributors to this volume have formed a community of scholars concerned with troubling dominant theories and paradigms in education and policymaking, particularly in terms of globalized discourses that construct the "normal" and "desirable."

Ranging from close readings of curricular and classroom discourse texts in several disciplines to cross-national critical policy studies, all chapters interrogate the notion of universal "best practice" or "best policies" for *all*. Consistent with the general themes and problematics of the new book series, this book opens space for more subtle and complex understandings of the child in the world, and adds to a postmodern critique of policies and practices that purport to be for "all," while excluding or providing different education for those who are constructed as different or deficient.

Beth Blue Swadener
Gaile S. Cannella
Marianne Bloch
Series Editors

References

Alvermann, D. E. (2000). "Researching Libraries, Literacies and Lives: A Rhizoanalysis." In E. St. Pierre and W. Pillow (Eds.), *Working the Ruins, Feminist Poststructural Theory and Methods in Education* (pp. 114–129). New Brunswick, NJ: Rutgers University Press.

Deleuze, Gilles & Guattari, Felix (1987). *A Thousand Plateaus: Capitalism and Schizophrenia* (trans. Brian Massumi). Minneapolis: University of Minnesota Press.

Foreword: Hopes of Inclusion/Recognition and Productions of Difference

This volume brings to the fore a particular set of intellectual questions and strategies for studying schooling. The studies ask the following questions albeit through a family of resemblances. How does it become possible to think and act as we do? How are the subjectivities through which we act and "think" produced historically? And, How do "thought" and reason function as practices that produce systems of inclusion and exclusion?

The concern with "thought" and reason is to rethink the politics of schooling. If I look at critical studies in social science and education, the problem of power is one that assumes a notion of sovereignty. The underlying question is how power is exercised to enable the rulers to rule and others to be ruled. Critiques of neoliberal policies are an example. The emphasis is on, for example, who benefits and is handicapped through school vouchers and market-oriented reforms. The end point of such research is locating the origin of power in order to provoke challenges and locate the origins of those that prevent change. In this critical tradition, studies of the family and child make the knowledge of schooling for examples, an epiphenomenon to transcendental social forces, or the reverse as a hermeneutic that makes context as natural, the voice of the "real" teacher as removed from history.

The current volume offers a complementary yet different notion of power. The concern is with the systems of reason through which the objects of reflection and participation are produced. My linking notions of reflection and participation here is not accidental. It is to recognize that knowledge is not merely about "ideas" that represent social or political interests of *the real*. A particular "fact" of modernity is that power is exercised less through brute force and more through systems of reason that order and classify what is known and acted on.[1] This notion of power is important today as exemplified most feminist studies, critical race theories and postcolonial scholarship. The rules and standards of reason are effects of power that generate principles about who we are and should be; and thus stand not against the real but are embodied in its productions. Therefore, the studies in this book can be read as challenging the adequacy of structural descriptions that do not take into account knowledge as an effect of power that orders and classifies objects of reflection and action in reforms or theories of the child and family.

The notion of the history of the present is central, a phrase used in multiple chapters of the book. This notion of history, or if I use the term "historicizing," is concerned with changes in the systems of reason in which objects are "seen," thought, and acted on to make the subject and orders who the self is and should be. The problem of change runs counter to a historicist focus on the temporal evolution of events defined through archival materials that stand as the positive, observable testimonies to the events of the past. The history of the present, in contrast, is an analysis of the politics of knowledge as a theory of change: to understand how the objects that order reflection and participation are formed and mutate over time; to consider how the systems of reason intern and enclose possibilities and produce inclusion and exclusion; and to denaturalize what is taken as natural and inevitable and thus to open up the possibilities of searching for other alternatives.

The comparative (historical as well as geographical) qualities of the studies in this book undo the provincialization of the modern school by placing its practices within changing patterns of globalization since at least the nineteenth century that circulate in Asia, Europe, and North and South America. If we use the school as a focal point, today's infatuations with globalization are inadequately understood and misrecognized without a historical perspective that examines what is distinctive from and similar to the past.

Before proceeding, my observations should be read as not without affection and affiliation. The studies in this book emerge as part of a broader program of theory and research within the Department of Curriculum and Instruction at the University of Wisconsin-Madison that has developed over the past few decades. Recognizing this collective location is not to claim a particularity and uniqueness as the research intersects with a broader interdisciplinary scholarship and dialogue that gives it its sustenance and place. The intellectual program, if I can call it a program without signifying intentionality but historicity, nonetheless, developed a working space within a diverse department to bring into play a critical dialogue about the politics of schooling.

In the remainder of this foreward, I focus briefly on three themes that circulate through the volume and provide contributions to the study of early childhood and elementary education, and of modern schooling. One theme is the education sciences as planning and designing society through designing the child. Second is the notion of standards as a governing practice. And third is the relation of the sublime and reason in the education sciences.

To Enact Changes In The Conditions Of People Also Enacts
Changes In The Kinds Of People That They Are And Should Be

There is an orthodoxy of education research about its being "useful" to change. That usefulness is to describe "what works" and its corollary, what is not working, so that better planning can be achieved to change conditions and, without stating its overlapping end, to change people. The legislative function of science qua planning is not new. The social sciences at the turn of the century were to provide knowledge that assisted the state's practices of social amelioration and reinstalling a moral order, particularly in the new populations of the city. Today's discussion of education science returns to the turn of the past century's phrase of "social engineering" but within a different assembly of ideas institutions and strategies. The sacredness of research as planning is implanted in distinctions that separate "theory" from "practice" and in talk about finding out "what really is happening" (context) versus what policy makers say (discourse/ text). The conservatives among us ask how we can plan so people will make the *right* choices. The left political activist quotes (and misquotes) Marx to implore that research(ers) serve as agents of change through generating practical knowledge for teachers to be emancipated.

But this is *not how it works*, to play with the current trope. The planning of people is not merely to educate the subject to make right choices or to emancipate and give voice of those previously excluded. From notions of health to the family and childhood explored in this book, pedagogy has been to make the modern "citizen" and society through constructing principles to order self-governing individuals. Science continues historically as systems of theorizing, observing, and classifying that open the innate qualities of humans for calculation and administration. Ordering "thought," problem-solving, and the rules of communication and community, pedagogy and its sciences embody principles to give stability and consensus to individual participation in a world deemed in flux.

If the planning of the child was only about self-governing and emancipation, then the question of concern would be only who has the better planning models or ideological claims. But it is not. The sciences of planning embody a particular comparative method associated with European modernity and the thought of the Enlightenment. Cassirer (1932/1951) argues, for example, that the crumbling of the classical and medieval conception of the "cosmos" was accompanied with Enlightenment's notions of the human mind and reason that measures, compares, combines, and differentiates the things in the world, including humanity's nature. Reason enables this comparison in a long and uneven history, and becomes an object of political intervention through human sciences, expressed in education in the language of children—how "to think," problem solve, and so on. Reason is a calculated subject and instrument for change itself, as pedagogy is to create the modern self and citizen through ordering children's reflection and action. The reason of the Enlightenment, however, was not only about the reasonable child/citizen. It embodied a comparative mode that placed human beings along a comparative continuum of values that divided the civilized from those who did not have the capabilities of reason.

Why dwell on this comparativeness of "reason" in modernity and its pedagogical projects? At one level, it is a way to think about the universal and the particular in discussions of the child and family. The theories of the social and education sciences, for example, embody continuums of values that differentiate the qualities and characteristics of the child who is or can become "civilized" from those who are placed in spaces outside of "reason" and its notions of "reasonable" people. There is little talk today about the civilized in pedagogy (versus what is found in larger political rhetoric of governmental regimes). Rather distinctions between the civilized and uncivilized are embodied in words like developmental norms and learning/learners that differentiate those who were not as advanced or uncivilized.

The "thought" that produces the comparative qualities of the universal and the particular has other qualities. Universal and particular qualities in pedagogy are embodied in comparative discourses that differentiate the hope of the future from the fears of collective dangers of those not properly schooled and of the dangerous populations that the school is to recognize for rescue. Narratives and images of hope, it can be argued, bring universalized cosmopolitan themes to theories as the qualities of the good, successful, and moral child and family. The hope of inclusion is comparative, with systems of recognition and difference of those who do not embody the universal qualities of the citizen. Schooling at the turn of the century, for example, embodied a cosmopolitan hope of the making of the future citizen from urban immigrants and racial groups through socialization processes of the school. The recognition of these populations and educational theories and programs also produced them as different. The narratives embodied fears about moral disorder of the city.

The hopes and fears of the city and its moral disorder continue in present reforms about urban renewal and urban education. The comparative qualities of pedagogical thought inscribed exclusion with its projects of inclusion through the transcendental values and norms of today's lifelong learner. That learner is different from the categorical child of "no child left behind."

Standards as Making the Child Legible and Manageable for Governing
A different contribution of the book is to locate the curriculum standards reforms within a historical accounting of the process of governing. Standards are governing practices of the modern state that can be traced in Europe in the seventeenth century. Standards were invented to develop the capacity to have direct knowledge and reliable means of enumerating and locating the population of the state in order to intervene and regulate the people of a realm. People had no last names to be put into the census to track. Measurement was almost random as each local area had its own system to measure (*a hand, a foot, a cartload, basketful, handful, within earshot*) that prevented any central administration (Scott, 1998). Standards were a way of mapping health and wealth of territories so the state knew who fell under its domain.

Standards were important to the notion of equality in the rise of the republics. The academicians of the revolutionary Republic of France, for example, saw the metric system as an intellectually important instrument to make the equal citizen. If the citizen did not have equal rights in relation to measurements, then it was assumed that the citizen might also have unequal rights in law and would pose the greatest obstacle to making a single people.

The creation of standards to have the equal citizen is embodied in contemporary discussions of educational standards. The discussions in this volume enable us to understand, for example, that the standards curriculum movements are governing practices to develop the right classification and the correct sorting devices for charting a course of action that will change society for the better and that will prevent the child being left behind from a future of joining in the ranks of those who deviate from the norm. But this hope of the future embodies standards that differentiate and divide through fabricating, borrowing from Hacking (1999), *kinds of people*. The "child left behind" is one human kind fabricated in the standards reform literature. It is a determinate category linked to issues of equity and inclusion. The category of "the child left behind" recognizes particular kinds of children as different and in need of rescue. That recognition of difference assumes a universal child spoken of with distinctions that are similar to those of the urban and at-risk child and with qualities different from but also to become like "all children who learn." The signifier of the "all" children functions as unspoken characteristics that normalize and differentiate the standards from which to recognize and divide the qualities of the rescued child.

A third theme is the sublime expressed in this volume through travel through discussion of the child's soul, themes of salvation and redemption in pedagogy. The concern with the sublime might seem misplaced as science is to bring disenchantment with the world. That disenchantment is to eliminate provincial values, local traditions, magic, and ideological sides in modernity. But the disenchantment is not all it seems. The disenchantment was continually doubled with enchantments in the enlightentant as rationality was overlaid with a sublime in which science embodied hope, awe, aesthetics, beauty, and fears. The relation of science to enchantments is itself a central theme in the post-Darwinist culture in the United States. From Augustus Comte's manifesto about positivism to William James' and Dewey's discussions about democracy, cultural theses overlapped the sublime with science, enchantments, and disenchantments. In addition, Max Weber (1904–1905/1958) argued that the social (and education) sciences brought to bear particular Puritan notions of salvation into a secular world.

The volume's historical and empirical analyses as well as its comparative approach are an intervention for rethinking the traditions through which to investigate the effects of schooling and its politics. The volume recognizes what Gaston Bachelard (1991) called "epistemological obstacles" that continue to circulate in contemporary studies of schooling. These obstacles are embodied in the distinction between nominalist (discourse, text) and realist (contexts). As is argued here and in the book's chapters, knowledge is not merely there to express the subject's intent but produces intent and purpose through the rules and standards of reason that generate principles of action. The bifurcated world of theory/experience and ideas/context obscures, divides, and erases traces of how expert knowledge works dialectically in the forming of social relations.

The results of the volume are not prescriptions for "change" as traditionally thought of, but about the importance of opening new spaces for conduct and action through critique of current reasoning. In doing so, it helps to reconsider issues of power and the politics of schooling.

Thomas S. Popkewitz
The University of Wisconsin-Madison

Note

1. I also recognize and appreciate the intellectual interventions of Giorgio Agamben (2003/2005) to rethink notions of sovereignty that enable political theory to deal with the construction of the subject and life.

References

Agamben, G. (2003–2005). *State of exception.* (K. Attell, Trans.). Chicago, IL: University of Chicago Press.

Bachelard, G. (1991). Epistemological Obstacles. In M. McAllester Jones (Ed.), *Gaston Bachelard subversive humanist: Texts and readings* (pp. 81–84). Madison: The University of Wisconsin Press.

Cassirer, E. (1932/1951). *The philosophy of the enlightenment* (F. Koelln & J. Pettegrove, Trans.). Princeton, NJ: Princeton University Press.

Hacking, I. (1986). Making up people. In T. C. Heller, M. Sosna, & D. E. Wellbery (Eds.), *Reconstructing individualism: Autonomy, individuality, and the self in Western thought* (pp. 222–236, 347–348). Stanford, CA: Stanford University Press.

Scott, J. (1998). *Seeing like a state: How certain schemes to improve the human condition have failed.* New Haven, CT: Yale University Press.

Weber, M. (1904–1905). *The Protestant ethic and the spirit of capitalism* (T. Parsons, Trans.). New York: Charles Scribner and Sons.

Acknowledgments

The editors of this book would like to thank the members of the University of Wisconsin-Madison "Thursday Group"—both local and international, for the many conversations that have inspired the book. We would particularly like to thank the following: Julie McLeod, Anne Lise Arneson, Stephen Thorpe, Andreas Fejes, Sabiha Bilgi, Miryam Espinosa Dulanto, and Barb Tarockoff.

This book has been stimulated by the frequent dialogues and relationships formed during the past years of the Reconceptualizing Early Childhood Conference (RECE), which had its inaugural meeting at the University of Wisconsin-Madison in October, 1991, and its most recent meeting also at UW–Madison in October, 2005. Indeed, several chapters in this volume were originally presented at the RECE conference held at Arizona State University.

We want to thank Beth Blue Swadener and Gaile Cannella for their editorial advice and the three series editors of the new Palgrave series on *Cultural Studies of Childhood* for allowing this volume to be published in it as one of the first volumes. We also acknowledge editorial help given by Amanda Johnson and Emily Leithauser from Palgrave-Macmillan Press during the publication process.

Finally, we would like to acknowledge many friends/families. Mimi Bloch thanks Peter and Ben Bloch, Emilie and Jesse Sondel, and, especially, Lore Bloch. Devorah Kennedy thanks her husband Mark Kennedy and her children Mai-Tal and Dotan Kennedy for reminding her that there is "life outside of her head," and her father Stanley Isenstein for his support of her family." She would also like to thank Annie Cook and Amy Sloane for helpful feedback. Dory Lightfoot acknowledges her husband Enrique Rueda-Sarmiento, her parents, Edwin and Lila Lightfoot for their constant support in her intellectual pursuits and her dogs Nube and Chiqui for sitting on her feet while she worked. Dar Weyenberg wishes to thank Kayla and Jamie Weyenberg for sharing their adventures with their horses Chief and Joey.

PART 1

The Child in the World

Introduction: Education and the Global/Local Construction of the Universal, Modern, and Globalized Child, School, and Nation

MARIANNE N. BLOCH, DEVORAH KENNEDY,
THEODORA LIGHTFOOT, AND DAR WEYENBERG

Educational discourses currently favored by governments across the globe focus attention on notions of accountability, standards, and best practices. These concepts, and the policies that promote them, are based on a common assumption—the assumption that it is possible to establish formulae through which to uniformly and objectively judge and assess all students. In the United States, the principles of standardization, accountability, and best practices are manifested in government policy such as *No Child Left Behind* directed at elementary education and beyond, and *Good Start/Grow Smart* directed at Head Start programs. Reforms and discourses that seem remarkably similar are evident in education initiatives appearing in various contexts, including the United Kingdom's Education Act 2002 and Taiwan's voucher program (both discussed in this volume). The similarity of educational discourses reflects common understandings of childhood and education. These include the concepts that (1) the child is knowable through scientific, objective study, (2) the knowledges derived through scientific study are universally applicable to all children, (3) educational practices can be derived from these knowledges and applied unilaterally to all children, and (4) that societal issues and problems can be addressed through educational processes designed to enhance individual development and learning.

Current politically based education initiatives are framed by reasoning through which childhood itself is understood as universal—that all children pass through similar stages and learn in similar ways. Thus, the child is understood as having particular characteristics, which curricula are to address in planning his/her education and preparing him/her to participate as a "good," "productive," and "educated" citizen, or community member, in the future. Too often, the values implicit in such concepts as childhood, best practices, education, and democratic citizenship remain unquestioned as they are implemented across the globe in differing cultural, economic, and political contexts.

Current reform initiatives stress the necessity of evidence-based educational and social science knowledge for framing pedagogy. While this emphasis on objective scientific knowledge applicable to all is not new, having existed in various forms since at least the 1960s, the need for a strong and cogent critique of the reasoning behind these ideas is urgent and timely. The supposedly universal notions of child development, early education policies, and high standards for "all" currently circulating cross-nationally, foster ideas that, because they are assumed to be applicable to all, they are assumed to be inclusive. Yet, the exclusions inscribed within the reasoning, and material effects of that reasoning, are overlooked as the knowledge is incorporated within current educational initiatives.

Contributors to this volume question the assumptions implicit within scientifically based universal knowledge of child development, standards, and best practices in education. Influenced by scholars from the Reconceptualizing Early Childhood Education group[1] (RECE), the authors challenge the idea that science-based knowledge related to the education of the young child is universally applicable, natural, or neutral. Contributions to this volume represent a response to the recent growth and intensification of efforts to privilege only scientifically based knowledge, and its assumption of universality, in the name of "fairness." This scholarly approach follows in the footsteps of many who initiated the use of critical theories in early and elementary education, critical psychological studies, and cultural studies of childhood to scrutinize current discursive languages and practices related to childhood, families, and education (e.g., Bloch, 1987, 1992; Burman, 1994; Cannella, 1997; James & Prout, 1997; Kessler & Swadener, 1992; Mallory & New, 1994; Pinar, 1997; Polakow, 1992, 1994; Prout *et al.*, 2001; Silin, 1985; Soto, 2000, 2002; Walkerdine, 1984, 1988, 1998; Walkerdine *et al.*, 2001). They also continue in the fashion of those who link childhood, families and education with local as well as global discourses of power and knowledge (e.g., Baker, 2001; Bloch & Popkewitz, 2000; Bloch *et al.*, 2003; Cannella & Viruru, 2004; Dahlberg *et al.*, 1999; Hultqvist & Dahlberg, 2001; Lesko, 2000; Tobin, 1999) building and advancing a critical tradition in the field of early and elementary education, and childhood studies, more generally. These writers ask questions such as the following: (1) How is it possible to "speak back" to universalizing concepts without becoming entrapped in simplistic answers and instrumental reform strategies?; (2) How can critical theories and methodologies in education be useful in this debate when definitions of "good" educational research are becoming increasingly narrow (see Shavelson & Towne, 2002; as examples of critique see, Bloch, 2004; Lather, 2004; Lincoln & Cannella, 2004; Popkewitz, 2004)?

The Education of Young Children as Rhizomatic Space

The conceptual approach for this chapter is Deleuze and Guattari's (1987) notion of "rhizomatic" space. Rhizomatic space incorporates three overlapping, complex, but distinct spaces. The first space is "striation" or "bounded space." This space denotes the linear, unitary, and progressive reasoning through which the object/subject of knowledge is constituted. In the case of early childhood studies, discourses on this level include the normative discourses of development and of universalizing "neutral" or "objective" concepts of the "good" child, the "good" parent or teacher, and of "best practices" in education. Another space referred to by Deleuze and Guattari as "lines of flight," is an undefined, abstract realm, where understandings and possibilities are so numerous that they cease to take on normative meaning (pp. 88–89). Lines of flight open up conceptual space and meanings attached to childhood and the possibility for new ways of "becoming" with a multiplicity of pathways that are not predefined or predetermined.

The chapters in this book are situated in an area between these spaces. This is the "rhizomatic" space, where it is possible to approach lines of flight and to question the boundaries produced in striated space. This is a space where we can reexamine our commonplace understandings and accepted truths without being so abstract or so open that our analysis has no connection and no meaning. In this space, texts, theories, and so on are not as important for "what" they are, as for how they relate to other texts, to other theories, and to other aspects of reality. As Deleuze and Guattari argue, "a rhizome ceaselessly establishes connections between semiotic chains, organizations of power, and circumstances . . . A semiotic chain is like a tuber, agglomerating very diverse acts." (1987, p. 7).

These complex connections encourage us to think outside of usual practices and to ask critical questions about the child and education as we know them. This allows us to hoist commonsensical notions of schooling practices out of the striated spaces rooted in enlightenment, colonializing, and modern knowledge, to deterritorialize what is currently accepted as reasonable and objective. Opening up of a multiplicity of spaces provides new opportunities for ways to think, act, and reason about present systems of thought, conduct, policy, and inclusions/exclusions.

The divergent concepts and strategies employed by the contributors provide a fertile field for interactions that produce a whole much more interesting than its parts. The chapters offer diverse and overlapping perspectives. When juxtaposed in the space of this book, they form possibilities for the creation of new meanings and new connections, producing a common yet varied project. The common project of both the book and the reconceptualizing tradition, is to showcase the way in which educational knowledge, which is commonly accepted as universally applicable, and which is often assumed to place us on a level playing field, at the same time creates an infinitude of ways to be abnormal, substandard, or aberrant. Standardized models of learning and development are instituted in a variety of settings such as schools, clinics, childcare centers, and the workplace. As Rose (1993) argues, by displaying diverse individuals in a common situation, schooling acts to make visible and to inscribe individual pathologies and irregularities with regards to a norm.

The international aspect of the book allows us to highlight the different forms educational and childhood discourses can take, and the diverse ways they interface, as they circulate internationally and across cultural/historical contexts. The authors come from a variety of cultures and national backgrounds including the United States, Japan, Taiwan, and Brazil, where the discourses of modern schooling and universal standards for the care and education of children circulate. Their analyses examine the effects of these discourses on constructions of difference, notions of underdevelopment, and modernization in these differing local contexts.

Approaches in Thinking About the Chapters

It is important to understand some of the underlying theoretical approaches that fall within the space represented by this book. The different approaches to the sustained questioning of accepted knowledge of the child and the education of the child that we have collected in this volume represent viewpoints and theoretical frameworks that are diverse enough to touch and interact with each other in many ways. At the same time, the theoretical approaches of the authors differ, offering varying ways to think about problems. The book also does not purport to replace one universal way of thinking with a single "new and better" one. Instead, it attempts to open multiple doors for new ways of thinking of childhood and education.

As such, the contributors employ a variety of different theoretical frameworks to raise questions about accepted, universalizing, and normalized understandings of childhood, family, education, as well as relations between global and local discourses of power and knowledge. Lee's chapter draws on Bourdieu's (1984) notions of cultural capital, habitus, and field in combination with Foucault's notions of power knowledge and governmentality (Foucault, 1980; 1991). Pauly's analysis uses Gramsci's (1971) and Hall's (e.g., 1989) critical cultural studies work as applied to visual cultures, media and education. Other authors (e.g., Bloch, Kennedy, Lightfoot, Qi, Weyenberg, and Peach) draw on Foucault's and Rose's (e.g., 1993, 1999) work to frame their arguments. Additionally, a range of feminist and postcolonial theorists (e.g., Bhabha, 1994; Chakrabarty, 2000; Escobar, 1995; Lather, 2004; Walkerdine, 1998; Walkerdine *et al.*, 2001; Young, 2000) are used to help frame and analyze identity construction, national imaginaries, and discourses that while situated within local and national contexts also travel globally. These disparate theoretical approaches offer readers a variety of ways to critique educational "universals," "objective standards," and "best practices."

However, situated within the rhizomatic space described above, there are underlying connections between the chapters in this volume. Although not all of the contributors explicitly draw on Michel Foucault's theoretical tradition, most of the chapters in this book can be seen in the light of his concept of "governmentality" (e.g., 1980, 1991).

Foucault's strategy of examining the relationships between power and knowledge in common discourses has made visible the extent to which knowledge which appears to be neutral is actually infused with power and is related to concepts of governing. The concept of "governmentality" refers to a way of looking at people as "populations," and of seeing those populations as resources to be mobilized, scrutinized, and used. As Foucault writes, governmentality constitutes a web of power.

> In the last years of the eighteenth century, European culture outlined a structure that has not yet been unraveled; we are only just beginning to disentangle a few of its threads, which are still so unknown to us that we immediately assume them to be either marvelously new or absolutely archaic, whereas for two hundred years (not less, yet not much more) they have constituted the dark, but firm web of our experience. (1963/1975, p. 199)

Foucault argues that these webs of experiences are still with us today and constitute who we are (also see Hacking, 1999). Many of the themes of the essays revolve around questions such as: What is the perceived "structure" of educational and child rearing experience? What are the contours of our conceptions of our experiences of schooling, childhood, parenting, education, health, family, and so on (e.g., Bloch et al., 2003)? What forms do these experiences take in daily practice? How have the educational experiences of the child been constructed historically, culturally, politically through power and knowledge relations (e.g., Hultqvist & Dahlberg, 2001; Stacey, 1996)?

Works written from the standpoint of understanding and questioning what we understand to be "normal" show the contingency of our taken-for-granted experiences and mechanisms of normalization. Mechanisms of normalization are crucial to the notion of "governmentality." As Foucault points out, since some time in the eighteenth century, the main purpose of governing has been to maximize the resources provided by a nation-state's population, and, in doing so, to increase the well being of individual members of that population (1991). Specifically, since the early decades of the nineteenth century, one of the primary foci of governmentality has been public schooling and education. Popkewitz (1998) describes governmentality as "the link between the modern state and the self-governing individual" (p. 77). The modern state, which holds some "guarantees of freedom" for its population, relies upon its citizens to function as self-disciplined individuals. Freedom relies upon acceptance of norms of conduct. Knowledge has been the link between notions of "normal" behavior and individual self-conduct. Power-knowledge relations reflect the construction of certain knowledge as truths that are historically and culturally located in relation to language or discourses imbued with power. It is these power-knowledge relationships that produce certain understandings and normalize conduct, thoughts and practices, while making alternative ways of acting or being seem "outside the norm," aberrant, different, less developed, or in need of modernization and development.

Education, be it theory, practice, policy or childrearing has functioned as a means of constituting individuals, inscribing notions of normalcy, and socially administering freedom. In the late nineteenth and early twentieth century, the scientific study of populations, observations of children, and categorization through assessments produced "new" types of "advanced" and "progressive" knowledge about childhood, the family, and schooling. This knowledge interacted with educational theory and practice to form dispositions of self-understanding and inform individual conduct. During this period, educators formed the assumption that it was possible and desirable to develop universally applicable "truths" about child development, about "proper" parenting, and about "scientifically based" teaching methodologies. Through education this knowledge was to form dispositions of self-understanding and inform individual conduct. This shift in the rules of reasoning—linking knowledge to the administration of freedom— particularly the production of what were assumed to be universally applicable truths about child development and scientifically based methods of teaching and parenting affected individuals understanding of themselves, others, and their conduct (Popkewitz & Bloch, 2001). Government became linked with self-governance. Strategies of surveillance, such as assessments or observations of children, age-stratified grading systems, charts and measurements at doctors' offices, or interventions into parenting, worked together to constitute "normal" conduct according to scientific, objective norms. At the same time, these concepts of "normal" made it possible to rank and order, include and exclude, in comparison to the norm; it also marked what was *not* normal development, learning, teaching, or childrearing.

In the early twentieth century, scholars became convinced that it was possible to construct standards that would apply to all and that it would be possible to test and evaluate standards and procedures for normal learning, high quality schooling, teaching, and parenting. It became possible to develop accepted technologies to measure and to categorize those who were normal, abnormal, and "gifted," and to align these designations with parenting, group membership, heredity, and teaching. We can see this type of reasoning both in early twentieth century technologies such as the development of IQ and achievement tests and in early twenty-first century concepts such as best practice or educational standards. Additionally, we see these ideas in legislation such as the *No Child Left Behind* and *Good Start, Grow Smart* policies that currently guide funding and practices in the United States, and globally in policies such as the Education Act 2002 in the United Kingdom and in voucher policies in Taiwan (see Lee and Peach, chapters 10 and 11, respectively, this volume).

However, while the discourses of the late twentieth and early twenty-first century seem similar to those of the late nineteenth and early twentieth century, they have shifted in subtle and significant ways as they (re)circulate and are translated in the context of different historical moments, and different local/global spaces. The social administration of what is normal now takes place within a globalized discourse of flexibility, autonomy, responsibility, and uncertainty—different from the discursive spaces of the early twentieth century that looked for certainty, and a stabilization of national and international contexts through scientific advancement, and an imagined democracy that would create progress and development (of the child and the nation) for all.

As Rose (1993) argues, the focus on universal standards, or on the use of universal criteria for analysis, are embedded in an unavoidable irony. Those educational experiences which appear to bring us together today, as they did in the early twentieth century, to assure quality, also differentiate and divide students in the process of uniting them. As they unite us, by placing us on a "fair" and "neutral" playing field, they also divide us, by highlighting and stigmatizing the ways in which children, families, parents, schools, teachers, classrooms, and communities differ. This ability to display people along a common plane combines, in social science, with a practice of reductionism, which tries to find the smallest number of factors that characterize humanity/normality and non-humanness/abnormality (Baker, 2003; Gardner, 1985). Human science discourse assumes that all students can be subsumed into a common space/place/identity and that special circumstance could be eliminated as distractions. In doing this, however, these ideas also produce an infinity of spaces where one can be different, abnormal, or aberrant, that can be collapsed into the closed space of at-risk, different, same, normally developed/abnormally developed with both conduct and reason defined and bounded, rather than open to different possibilities. All of the contributors to this volume explore the boundaries of these closed spaces.

In this volume, we are particularly concerned about shifting concepts of what is normal for particular ages as well as normal performance and conduct associated with notions of being productive, flexible, cooperative, and self-governing future citizens at a young age. We therefore focus on the irony and tension inherent in the way today's standards of normalcy always contain the potential for less than normal, or even more than normal, and that good citizenship or group membership implies multiple ways of belonging to a greater or lesser degree. There is a need to worry about the way these discourses close down possibilities rather than opening up spaces for the myriad of ways in which conduct, action, and identity can be thought of as well as performed.

Along with Foucault and Rose, who explore education from the standpoint of governmentality, many of our writers draw on other critiques of educational universals, including theorists such as Pierre Bourdieu, Homi Bhabha, and Dipesh Chakrabarty. Bhabha (1994) sets forth a number of ideas that can be used to explore the complicated relationship between universalizing knowledges and local cultures. His work helps us analyze discourses about educational practices, pedagogies, and policies or reforms in relation to both globalized discourses and hybridized and situated contingent and local practices. Bhabha's (1994) approach reminds us of the fluidity and contingency of foundational universals as well as the impact and importance of everyday life or performativity within a fluid process that is not to be confused with being easy or smooth. Finally, Chakrabarty (2000) uses a slightly different strategy to make a similar point—that the "same" is not really the same when experienced by people from different backgrounds.

The contributors to this book employ a variety of approaches to examine the relationship of the child in the world. Each chapter, questions and challenges the way in which theory, policy, or practice make visible, in microscopic detail, the unlimited ways in which it is possible to be backwards, delayed, primitive, or abnormal, in short to not be modern, civilized, normal. Similarly, in different contributions, the gendered, racialized, or linguistic discourses embedded in ideals of norms, universals, and standards are questioned along with the notion of what it means to be essentialized into the "other" who is or is not considered to be more or less rational, more or less emotional, more or less dependent and immature, and historically more or less "developed." While these descriptions can be taken to be those assigned to "men" or to "women," they have also been ascribed to adults/adolescents and to very young children (Cannella, 1997; Dahlberg *et al.*, 1999; Leavitt, 1994), to students from Euro-American or Western European "heritage" or to those considered to be different through descriptions such as "culturally diverse backgrounds" or "second language learners," or to those considered having dis/ability. As one example, Baker (2003) illustrates how deaf children are considered less than human in some literature about disability. The notion of disability cultures—*whole* cultures—ascribed to be disabled in *every* way reminds us of the extent to which modernity and the notion of modern, progressive rational *hearing* and thinking man defines normality, and, in so doing, abnormality. The current volume, while focusing particularly on schooling and education for younger children, illustrates a diversity of critical issues that can be elaborated and extended into schooling and education for different ages.

Topics in the book are diverse and involve an intersection of theoretical framing of problematics with an investigation of a particular topic in schooling, pedagogy, or policy related to childhood, family, and education. They are located in settings such as the "universal" kindergartens of Froebel or educational programs promoted by John Dewey. They are located in Jewish preschools that universalize while they appear to be particular. They are within universal pedagogies of language acquisition or in the self-disciplining of children in Japan and the "model" minority ascribed to Asian Americans. The topics in pedagogy range from the universal to the particulars of child development as a pedagogical base for instruction to chapters that focus attention on health education "for all," language instruction for the universal child, and policies for diverse countries—including contemporary pedagogical policies in the United Kingdom, Taiwan, Brazil, and Japan—that appear to be universally good because they are modern, progressive, and adopted from the "West," while at the same time marking others as exotic, less developed, or in need of modernization. These discursive practices exclude many from normality and/or divide the particular and different from the "universal" that is supposed to be applicable to "all."

The theoretical framing of the chapters vary, although the majority draw on different "post" (postmodern/poststructural/postcolonial) perspectives. The contributors examine the ways understandings of children create power-laden hybrid, multiple, and split identities. Some chapters challenge and cross the boundaries between and among postmodern, postcolonial, and feminist theories. The chapters represent varied methodologies. The techniques used for analysis range from textual/literary analysis to cultural historical analyses to discursive analysis of classroom observations, curriculum, and reform texts. What all the chapters share is an emphasis on the "close reading" of texts (whether from reform reports, media reports, examinations of architecture, assessments, schedules, groupings, or narrative reports of what people say). These approaches reflect critical feminist, poststructural, and postcolonial research traditions questioning the specifics of language or text, provide clues to important social meanings, and reflect power-knowledge relations.

Bourdieu (1984) points out that not only nationalities and ethnic groups but also groups within single societies do not have equal access to the knowledge, the experiences, the tastes, desires, and so on which lead to success in schooling, and careers later in life. Bourdieu's notion of reflexive sociology makes us aware that early childhood and elementary education appears to occur on a "playing field" where individual practice is guided by the habitus and capital of individuals, groups, and institutions. His ideas make it possible to examine the relational shifting of position with/in the global and local fields of childhood, education, and educational reforms as well as policies. These theorists have led the contributors to ask questions such as the following:

What are the effects of daily practice? What are the conditions that enable children to have one set of experiences and not others, to express desire/pleasures, and to construct themselves and others in certain ways? In what ways have constructions of the child as universal, developing, and normal affected teachers' practices or parents' and community perceptions of children as well as normal childhood? How have these conditions arranged the lives of children through particular schooling practices, including the architecture of the school, its daily schedule, and constructions of normal interactions, reasoning, and conduct? How do discourses shape what is possible and therefore not possible to think and act? How have groups arrived at particular sets of practices, or desired experiences, in current timespaces? How might rhizomatic understandings of openings, multiplicities of understandings and possibilities present different ways of acting and thinking?

Framing "The Child in the World/
The World in the Child" as a Volume: Contributions
Toward an Opening Up of Spaces/places

The contributors to this volume focus their analysis on education and schooling in relation to specific constructions of childhood, family, nation, and world. While the contributors examine childhood and education at the early and elementary educational levels, they use different national and curricular contexts to examine how the universal and modern child, family, and education have been configured, historically and culturally, as part of a globalized discourse of modernity within power and knowledge relations. The international aspect of this book offers readers a variety of viewpoints upon how circulation of normalizing educational discourses produce multiple hybrid variations of apparently uniform situations, methodologies, or goals. Internationally, the spread of the discourses of normalization, modernization, and development bring with them the same irony as we have seen with national standards. As standards move across the globe and hybridize in different locations, they highlight our differences even as they attempt to bring us to the same standards.

We have introduced this book as an assemblage of varying examinations and critiques of early childhood and childhood education. This enables the contributors to deploy divergent tactics in "deterritorializing" (see Deleuze & Guattari, 1987) accepted notions of early childhood and elementary education. All the authors explore the effects of the power-knowledge relationships that create and limit the child/parent/teacher as subjects. They trouble accepted ideas and scientific beliefs by exploring the manner in which these ideas and beliefs "govern" normal conduct.

In organizing the layout of this book, we have divided it into the usual divisions of sections and chapters. However, the chapters are not to be seen in an ordinary sense of providing a linear sense of organization that goes neatly from first to last. Although the chapters initially may seem like a linear progression from one chapter to the next chapter, in reality there are no real beginnings or ends. The value of using the notion of a rhizome is that there is no such notion as a first, middle, or last chapter. Each chapter can be read independently of others, and the chapters can be reorganized and juxtaposed in various ways, causing different relationships to appear between them. The chapters are "regions of intensity" (Deleuze & Guattari, 1987, p. 22). Each is inscribed as a historical marker providing an account of a fissure into commonsense ways of thinking of the child. Each is a "multiplicity connected to other multiplicities" whereby the reader can focus on a particular story but yet come to realize other histories exist along side of a particular history. In this sense, each chapter is a fragment in the histories of our present.

This part of the volume, *"Education and the Global/Local Construction of the Universal, Modern and Globalized Child, School and Nation"* sets the stage for the chapters in the second, third, and fourth sections. Chapters in the second part, "*Governing the Universal Modern Child and Family*," examine the cultural discourses of national as well as ethnic narratives that construct the notion of the universal modern child and the idealized and universal "good" family, defined as a populational category from which others (groups, families, children) are measured against. These chapters represent illustrations of the complexities and the exclusionary reasoning embedded in modern discourses about childhood and the family. By complicating simple notions of the family, child story, inclusions as well as exclusions are opened up for scrutiny; new possibilities are envisioned. In chapter 1, Bloch explores "foundational" linkages between childhood, education, and democracy through a cultural historical analysis of texts written by educational philosophers/theorists/scientists at the end of the nineteenth and beginning of the twentieth centuries. Employing Foucault's notions of governmentality and power-knowledge, Bloch explores shifts in reasoning that produces and reinscribes notions of normal and abnormal, particularly within early education pedagogical practices.

Chapter 2 also addresses differentiations emerging through reasoning about "normal" childhood. Kennedy draws upon postcolonial theories of Bhabha and Chakrabarty, to examine inscriptions of difference as discourses of high quality child care are translated into practice within a Jewish preschool setting. Adler, in chapter 3, also employs Bhabha's approach to cultural difference. Adler reflects upon problems of Asian American families in negotiating "Western" and "Eastern" conceptions of "good" school behavior. The negotiation is complicated by the construction of Asian Americans as a homogeneous "model minority." Adler examines the diversity of ethnicities, nationalities, and cultures that make up this category and their differing negotiation of insider/outsider status.

In part 3, "*Governing the Modern and Normal Child Through Pedagogical Discourses*," the notion of universality is examined through looking at the spread of pedagogical ideas around the world, the global discourses of universality as these translate and travel across space and places of childhood, family, and schooling, and as they come into geographical and conceptual spaces differently.

In chapter 4, Lightfoot looks at the intersection between concepts of first and second language acquisition, and socially and historically contingent concepts of the "productive" citizen. She argues first language acquisition theory, cannot ever be neutral, or separate from the social, economic, intellectual, and political culture that linguists and cognitive theorists are living in. Second, she looks at some of the dangers involved in transferring theories about how young children acquire their first language formulated in the concept of postindustrial norms to second language students from a wide variety of historical and cultural backgrounds.

Weyenberg historicizes current notions of health promotion, especially as it relates to the formation of subjectivities as effects of power-knowledge. The author explores, through a history of the present, how medical or scientific knowledge functions within health discourses to fashion particular kinds of individuals. In chapter 5, Weyenberg also examines how this "knowledge" of the individual was articulated over time in terms of historical normative constructions related to how one should conduct one's life in order to maintain health through a gendered curriculum. In chapter 6, Pauly uses theories drawn from the field of visual culture studies to look at the ways the media constructs young children as universal consumers. She then examines the work of a group of teacher education students who have used postmodern theory to deconstruct the media images that shaped their own childhood.

Qi, in chapter 7, explores how various technologies have constructed the notion of childhood as a way of disciplining and self-disciplining in contemporary Japan. She explores disciplinary power and normalization in schooling practices and argues that the construction of the young child involves complex power relations. The effect of these discourses is that Japanese children are expected to become self-governing and thus to "enjoy" controlling their own behavior. The eighth chapter is an analysis of Brazilian educational texts. Lima looks at how it became possible to think about children's problems in a managerial, or, referring to Foucault, a "governmental" sense. She argues that concerns with pupils' maladjustments were related to the expansion of the educational system and the entrance into it of a new group of children whose parents had never had the opportunity to attend school. These children were seen as "different," and "strange," even if they could not be properly designated to be "abnormal." Once the idea of "problem children" was raised, all children became "at risk" of becoming "problems," through a large and diverse set of circumstances such as divorce, the birth of a sibling, or even entrance into puberty.

The fourth part, "*Governing the Modern and Post-modern Citizen and Nation Through Universal Reforms in Education*" focuses on examples of policy and pedagogy in the United Kingdom, Taiwan and the United States. Pena, in chapter 9, deals with some of the ways that religious patterns of thought haunt the thinking of the educational legislation, *No Child Left Behind*, and is grounded in the work of Foucault. It seeks to differentiate styles of religious thought along a historical continuum to demonstrate how these patterns translate into present ways of conceptualizing students as the same or conceptualizing them as difference. The essay interrogates the present understandings of the vocabulary of the legislation and contrasts these meanings with other historical understandings of these "words" with particular attention to their religious associations.

Drawing from Foucault's notion of governmentality, Lee, in Chapter 10, theorizes that early childhood educational vouchers create possibilities for parents to embody different identities as they internalize notions of "choice" in education. Conceptualizing educational change and reform as "anthropological phenomena," this chapter raises questions about how educational vouchers travel as a form of an "indigenous foreigner" to take their current form in Taiwan, reflecting a form of hybridity in discourses from global to local levels. Peach, in chapter 11, looks at an educational policy shift that creates a new type of child, the "Foundation Stage" child as an important educational problematic. This new child, consisting of three- to five-year-olds, is situated by this policy at the beginning of primary school. Foundation stage children are highlighted as a new resource for the national state as human capital while their education and care is shaped by the marketization of education

Note

1. The Reconceptualizing group is an internationally based group of scholars who come together on a yearly basis to discuss early childhood education from multiple perspectives. Employing perspectives including postmodernism, critical theory, and poststructural feminism the scholars examine assumptions within current early childhood discourse and policy initiatives. In so doing, they bring to light inequities inscribed within the knowledge.

References

Baker, B. (2001). *In perpetual motion: Theories of power, educational history, and the child.* New York: Peter Lang.

Baker, B. (2003). Hear ye! Hear ye! Language, deaf education, and the governance of the child in historical perspective. In M. Bloch, K. Holmlund, I. Moqvist and T. S. Popkewitz (Eds.), *Governing children, families, and education: Restructuring the welfare state* (pp. 287–312). New York: Palgrave Macmillan.

Bhabha, H. (1994). *The location of culture.* London: Routledge. Bloch, M. N. (1987). Becoming scientific and professional: Historical perspectives on early childhood education and child care. In T. S. Popkewitz (Ed.), *The formation of school subjects* (pp. 25–62). Philadelphia, PA: Falmer.

Bloch, M. N. (1992). Critical science and the relationship between child development research and early education. In S. Kessler & E. B. Swadener (Eds.), *Reconceptualizing early education curriculum: Beginning the dialogue.* New York: Teachers College Press.

Bloch, M. N. (2004). A discourse that disciplines, governs, and regulates: The National Research Council's Report on Scientific Research in Education. *Qualitative Inquiry, 10* (1), 96–110.

Bloch, M. N. & Popkewitz, T. S. (2000). Constructing the child, parent, and teacher: Discourses on development. In L. D. Soto (Ed.), *The politics of early childhood education.* New York: Peter Lang.

Bloch, M. N., Holmlund, K., Moqvist, I., & Popkewitz, T. S. (Eds.) (2003). *Governing children, families, and education: Restructuring the welfare state.* New York: Palgrave.

Bourdieu, P. (1984). *Distinction: A social critique of the judgment of taste.* Cambridge, MA: Harvard University Press.

Burman, E. (1994). *Deconstructing developmental psychology.* New York: Routledge.

Cannella, G. S. (1997). *Deconstructing early childhood education: Social justice & revolution.* New York: Peter Lang.

Cannella, G. & Viruru, R. (2004). *Childhood and postcolonization.* Philadelphia, PA: Routledge Falmer.

Chakrabarty, D. (2000). *Provincializing Europe: Postcolonial thought and historical difference.* Princeton, NJ: Princeton University Press.

Dahlberg, G., Moss, P., & Pence, A. (1999). *Beyond quality in early childhood education: Postmodern perspectives.* London: Routledge Press.

Deleuze, G., & Guattari, F. (1987). *A thousand plateaus: Capitalism and schizophrenia* (B. Massumi, Trans.). Minneapolis: University of Minnesota Press.

Escobar, A. (1995). *Encountering development.* Princeton, NJ: Princeton University Press.

Foucault, M. (1975). *The birth of the clinic: An archaeology of medical perception*. New York: Vintage Books. (Original work published 1963.)

Foucault, M. (1980). *Power/knowledge: Selected interviews and writings, 1972–1977*. (Edited by Colin Gordon.) New York: Pantheon Books.

Foucault, M. (1991). Governmentality. In G. Burchell, C. Gordon & P. Miller (Eds.) *The Foucault effect: Studies in governmentality* (pp. 87–104). Chicago, IL: University of Chicago Press.

Gardner, H. (1985). *Mind's new science: A history of the cognitive revolution*. New York: Basic Books.

Gramsci, A. (1971). *Selections from the prison notebooks*. International Pub: New York.

Hacking, I. (1999). *The social construction of what?* Cambridge, MA: Harvard University Press.

Hall, S. (1989). Ethnicity: Identities and difference. *Radical America, 23*(4), 9–20.

Hultqvist, K., & Dahlberg, G. (Eds.) (2001). *Governing the child in the new millennium*. London: Routledge Press.

James, A. & Prout, A. (Eds.) (1997). *Constructing and reconstructing childhood: Contemporary issues in the sociological study of childhood*. London: Falmer.

Kessler, S., & Swadener, E. B. (1992). *Reconceptualizing the early childhood curriculum: Beginning the dialogue*. New York: Teachers College Press.

Lather, P. (2004). This is your father's paradigm: Government intrusion and the case of qualitative research in education. *Qualitative Inquiry, 10* (1), 15–34.

Leavitt, R. L. (1994). *Power and emotion in infant-toddler day care*. New York: State University of New York Press.

Lesko, N. (2000). *Act your age*. New York: Routledge.

Lincoln, Y. S., & Cannella G. S. (2004). Dangerous discourses: Methodological conservatism and governmental regimes of truth. *Qualitative Inquiry, 10* (1), 5–14.

Mallory, B., & New, R. (Eds.) (1994). *Cultural diversity and developmentally appropriate practices*. New York: Teachers College Press.

Pinar, W. (1997). The reconceptionalization of curriculum studies. In D. J. Flinders & S. J. Thornton, *The curriculum studies reader* (pp.121–129). New York: Routledge.

Polakow, V. (1992). *The erosion of childhood* (2nd ed.). Chicago, IL: University of Chicago Press.

Polakow, V. (1994). *Lives on the edge: Single mothers and their children in the other America*. Chicago, IL: University of Chicago Press.

Popkewitz, T. (1998). Dewey, Vygotsky, and the social administration of the individual: Constructivist pedagogy as systems of ideas in historical spaces. *American Educational Research Journal, 35* (4), 535–570.

Popkewitz, T. S. (2004). Is the National Research Council Committee's Report on Scientific Research in Education scientific? On trusting the manifesto. *Qualitative Inquiry, 10* (1), 79–95.

Popkewitz, T. S., & Bloch, M. N. (2001). Administering freedom: A history of the present—rescuing the parent to rescue the child for society. In K. Hultqvist, & G. Dahlberg (Eds.). *Governing the child in the new millennium* (pp. 85–118). London: Routledge.

Prout, A., Jenks, C., & James, A. (2001). *Theorizing childhood*. New York: Teachers College Press.

Rose, N. (1993). *Governing the soul: The shaping of the private self*. New York: Routledge.

Rose, N. (1999). *Powers of freedom: Reframing political thought*. Cambridge: Cambridge University Press.

Shavelson, R., & Towne, L. (Eds.) (2002). *Good scientific research in education*. Washington, DC: National Research Council.

Silin, J. G. (1985). The early childhood educator's knowledge base: A reconsideration. In L. Katz (Ed.) *Advances in research in education*. Norwood, NJ: Ablex Publishing.

Soto, L. D. (2000). *The politics of early childhood education*. New York: Peter Lang.

Soto, L. D. (2002). *Making a difference in the lives of bilingual/bicultural children*. New York: Peter Lang.

Stacey, J. (1996). *In the name of the family: Rethinking family values in the postmodern age*. Boston, MA: Beacon Press.

Tobin, J. (Ed.) (1999). *Making a place for pleasure*. New Haven, CT: Yale University Press.

Walkerdine, V. (1984). Developmental psychology and the child-centered pedagogy: The insertion of Piaget into early education. In J. Henriques, W. Hollway, C. Udwin, C. Vern & V. Walkerdine (Eds.), *Changing the subject: Psychology, social regulation, and subjectivity* (pp. 153–177). London: Routledge.

Walkerdine, V. (1988). *The mastery of reason: Cognitive development and the production of rationality*. London: Routledge Press.

Walkerdine, V. (1998). *Daddy's girl: Young girls and popular culture.* Cambridge: Harvard University Press.

Walkerdine, V., Lucey, H., & Melody, J. (2001). *Growing up girl: Psychosocial explorations of gender and class.* New York: New York University Press.

Young, R. J. C. (2000). Deconstruction and the postcolonial. In N. Royle (Ed.). *Deconstructions: A user's guide.* Houndmills: Palgrave.

Governing the Universal, Modern Child and Family

CHAPTER 1

Educational Theories and Pedagogies as Technologies of Power/Knowledge: Educating the Young Child as a Citizen of an Imagined Nation and World

MARIANNE N. BLOCH[*]

Past and Present Reasoning

This chapter uses a cultural historical approach to focus on early childhood programs in the late nineteenth and early twentieth centuries in the United States, and new discursive languages and practices in the early twenty-first century. During the first period, new discourses came to govern the reason through which children, parents, teachers, and programs were constructed as modern, well-educated, developed and civilized (Bloch & Popkewitz, 2000). While I draw on two enlightenment philosophers' ideas, Rousseau's and Locke's, as influences on nineteenth century discourses, transcendental, idealist, and evolutionary social biological theories also framed early childhood pedagogical practices as these were formulated in the nineteenth century in Europe and the United States. These discourses are also part of a larger assemblage of constructed imaginaries about which children, parents, teachers, and programs were good, and, by contrast, were also bad, and, therefore, in need of intervention to become better, assimilated citizens in the changing climate of the United States at the turn of the century.

By the early twentieth century, new discourses emerged that focused ever greater attention on the modern, scientific, democratic and progressive child, as well as on the ways in which a child's conduct and habits could be tempered through behavioral (Hill, Thorndike), progressive (Addams, Dewey, Mitchell, Pratt), or therapeutic (Freud, Erikson) pedagogical and psychological teaching and childrearing environments (see Bloch, 1987). As the constructions of childhood and their relation to different forms of education played out in the early twentieth century, a focus on children's biological, emotional, linguistic, and social *development* merged with the conception of a findable set of universal truths and predictable laws of child development that guide teachers and parents, doctors, and social workers in their work to characterize normal and "scientifically" knowable (observable, testable) children. Currently, hard scientific evidence is again called for to determine which children are successes and which ones are falling behind (e.g., the *No Child Left Behind* reform). The call for scientifically rigorous "evidence" based findings and experimental research, and the testable and observable child, family, and teacher has expanded into the investigation of the neurons of the brain and into the biological make-up of the "ADHD," not so docile body of young children.

Welfare policies in the United States focus enormous attention on how early education can enhance productive citizenship-training, as well as how preschooling/kindergarten can assimilate children, perceived as different by class, race, language, ability/disability, to the norms expected of the nation. World Bank and other international agency documents globalize the idea that preschool education may prove to be a critical reform for early childhood development and for the production of future national, *modern, developed, and educated* citizens (Bloch, 2003). These discourses carry within them reasoning about *all* children's development and education, and the universal desire to be modern. But the reasoning that appears inclusive of all also carries many ideas, identities, and patterns of conduct that are exclusionary (also see Bloch *et al.*, 2003).

The discourses of the past are not the same as the present, there are certain continuities and ruptures such that it becomes important to look at each period and place to understand how the circulation of discursive languages and practices occurs across nations as well as how these "settle" and take on shape within different localities. Thus, while today new discourses (languages, practices, reforms) appear to travel across borders quickly, they are translated differently. Whether urban citizenship, education for immigrant children, national or global citizenship is debated, how education and schooling are imagined to define how we think and conduct ourselves and imagine others embed philosophies of time and citizenship, born in modern thinking, but now enmeshed in "postmodern," culturally uncertain times. The pedagogical possibilities are also open, in need of constant deconstruction, critique, and reconceptualization. Drawing on Deleuze and Guattari's (1987) concept of the rhizome, these are moments full of danger, and new possibilities, opening up new spaces for thought and action.

The ways of reasoning that were fabricated from a complex amalgamation of different discursive languages and practices circulated broadly in the nineteenth and twentieth centuries, as different ways of reasoning circulate today. In the past, and the present, the assemblage of discourses have influenced how we form our own subjectivities as well as how individuals, groups, and nations, think about themselves and others. The cultural reasoning systems were and are related to *power/knowledge* relations, how we were and are governed to think about truth, who had (and has) the authority to speak, how we came (and come) to define what was/is good for children, or what we now term standards of "best practice" (Foucault, 1980, p. 131). In the second section of the chapter, I discuss the concept of citizenship as a *national imaginary*, and then I briefly discuss Foucault's notions of a *history of the present* and *governmentality* as these are used to examine discourses that construct reasoning about the young child.

The third section of the chapter uses selected primary and secondary sources to reflect on the young child imagined as future citizen in Friedrich Froebel's new *kindergarten* program in Germany and then as it was translated into the United States in the last half of the nineteenth century. I then turn to John Dewey's subprimary program as well as selections from Jane Addams' ideas from her work at the Chicago Hull House in the early twentieth century that also reflect on the young child as (future) democratic citizen. These texts are used to examine how images or *imaginaries* of the relationship between early education, child care, and democratic education of young children and their families as future citizens of the United States are presented and embed rationalities of citizenship, inclusion, and exclusion in reasoning (see also Popkewitz & Bloch, 2001). In the last section of the chapter, the systems of reasoning that permeated ideas at the beginning of the twentieth century are contrasted with discourses on the young child and curriculum in the twenty-first century.

Global-Local Imaginations of the Child as Future Citizen

Traveling Discourses and the Translation of
Knowledge across Nations

In *Kindergarten and cultures: The global diffusion of an idea*, Wollons (2002) introduces the idea of Froebelian kindergartens' diffusion to diverse countries during the latter part of the nineteenth century. She speaks of "the kindergarten as a politicized institution, directly linked to the goals of the state in the formation of national identity, citizenship, and moral values" (p. 2). Wollons discusses the translation of ideas across nations and their interaction with indigenous cultural values, and different historical, political, and cultural systems of reasoning about citizenship. This introduces the idea that discursive practices are contingent on the local geographical context in the reception, *translation of, and indigenous acceptance of* global ideas about new institutions (e.g., the kindergarten) (Bhabha, 1994; Bloch, 2003; and Popkewitz, 2000 on global/local discourses).

The notion of translation is not simple or unidirectional when speaking about the conveyance of ideas from one cultural context to another; indeed ideas are likely to circulate and settle in different locations in complex ways, and then return to a global "stage" of discussion as new ideas circulate around the globe (Bloch, 2003, Popkewitz, 2000). Notions of modern schooling, in general, including early childhood education, that increased in popularity and importance around the world during the nineteenth and twentieth centuries, are good case examples. O'Malley (1998) suggests:

> translation implies a process in which (state) programmers "make sense" of the indigenous governances—ignoring aspects which are "incomprehensible," thinking of practices as if they were situated within a familiar rather than an alien culture, "correcting" obvious "errors," assigning significance according to familiar rather than to alien priorities (p. 162).

The translation of new educational ideas, such as the Froebelian kindergarten, from Europe to America, therefore, is complex. It leads to a discussion of imaginations of citizenship, nations, and communities.

An Imagined Community

Anderson's *Imagined Community* (1983/1991) illustrates the constructed notion of nation, and imagined citizen that is an important guide for analysis of educational philosophies and practices in the eighteenth to twenty-first centuries.

> In an anthropological spirit, then, I propose the following definition of the *nation*: it is an imagined political community—and imagined as both inherently limited and *sovereign* . . . Finally, it (the nation) is imagined as a *community*, because, regardless of the actual inequality and exploitation that may prevail in each, the nation is always conceived as a deep, horizontal comradeship. (pp. 5–7, italics added)

Homi Bhabha (1990) also speaks of the "nation" as "narration" using postcolonial theory to critique the representations of cultures as homogeneous, or the "*pre-given* ethnic or cultural traits" (Bhabha, 1994, p. 2) that also inscribe assumptions of sameness/difference in imaginations of nation, culture, ethnic group, and community. Chakrabarty (2000), Said (1978), and Young (1995) point to the ways in which imaginaries of authority and civilization have been used to authorize conceptions of individuals, groups, or nations as different, underdeveloped, or uncivilized, and in need of intervention to become normal, developed, and civilized. As we think of education for citizenship as an imaginary, then we must also look at constructions of sameness/difference as related to social and historical articulations related to narratives of truth, power, and authority.

A History of the Present

Here I use a notion of cultural history (see Popkewitz *et al.*, 2001 for a larger discussion) to examine discursively organized patterns of reasoning that characterize a period, producing meanings within that moment. This notion of history is neither linearly nor causally connected with the present. A *History of the Present* illuminates taken-for-granted notions, such as "good child development" through a presentation of discursive patterns in the present, as well as in the past, illustrating how reason, knowledge, and truths are formed in certain timespaces. In using a history of the present, I also assume both continuities and ruptures between past and present reasoning. Nikolas Rose (1999) suggests, for example:

> Historical essays are . . . to disturb that which forms the very groundwork of our present, to make the given once more strange and to cause us to wonder at how it came to appear so natural. How have we been made up as governable subjects? (pp. 58)

In the next section, I use a history of the present, as well as the concept of governmentality (Foucault, 1991) to examine early childhood education by looking at continuities and ruptures in reasoning and governing from the eighteenth to twenty-first centuries.

Educating Democratic *Future* Citizens

We often take for granted the notion of development as evolutionary/biological, social, economic, and political depending upon whether we are reasoning about the individual, a group, the nation, or the world. A political/economic philosophy of educating toward the "future," linear development *over time*, an aspiration to make progress, given new information, often gained through scientific developments is also embedded in modern rationalities of schooling. These combine with an evolutionary idea of development that fabricates subjectivities about which individuals and nations are making "progress," are modern, or are developed, and which are not. In this section I examine these rationalities by focusing on selected ideas of Friedrich Froebel, John Dewey, and Jane Adams as early representatives of German and US-based educational reformers. I begin with a brief discussion of Rousseau's and Locke's work, who, among others, are interpreted as having discursively framed what could be envisaged in fabricating the early educational pedagogies in the nineteenth and twentieth century.

Rousseau, Locke, and the New Liberal, Rational Citizen

Early Childhood as a Stage of Childhood and Citizenship Development
Jean Jacques Rousseau's philosophies are often interpreted in terms of children being best left to themselves in nature rather than schoolrooms or with tutors/parents and learning by natural consequences (see *Emile*, by Rousseau, 1762/1979). His ideas were also interpreted as a critique of eighteenth century childrearing, pedagogical practices by tutors for elite

children, as well as of governing in France (Rousseau, 1762/1968). Rousseau's *Emile* was interpreted as encouraging (boys) to have an education toward autonomous, rational, and participatory and reasoning democratic citizenry, once the monarchy was overthrown; girls were to be trained to be good wives, dependent on their husbands' judgment and provisions, and to be prepared to train the next generation of citizens well (see Book V in Rousseau, 1762/1979).

John Locke's *Some thoughts concerning education* (Locke, 1693/1999) is similarly interpreted as promoting the idea of young children as blank slates (*tabula rasa*), individuals with different interests based upon experience, best taught through play, concrete games, and self-motivating activities. His treatises on government (Locke, 1689/1970) are the basis for liberal rationalities that emphasize autonomy, liberal individualism, individual responsibility, civil liberties and freedoms, and an educated participatory citizenship among elite males that counterbalances a monarch's rule.

These rationalities of governing privileged *unalienable* rights and obligations of citizenship, influencing the French, English, and Americans during the seventeenth and eighteenth centuries. They also influenced other philosophies about new modes of governing, including those held by German citizens during the (failed) "Revolution of 1848," after which some left Germany for the United States, and elsewhere, bringing ideas of the Froebelian kindergarten with them.

In the United States, the historical fear of a ruling executive's power over the individual resulted in the eighteenth century congress, favoring a form of liberalism that privileged individual autonomy, responsibility, civic and state rights as a balance to a strong executive or the monarchy they had experienced prior to the revolution. The philosophies of liberal governance embedded separations of state governing and a privileging of an individual's own care of himself and family or community. The focus on an autonomous, responsible and free individualism (for property-owning males) also embedded reasoning of the private family that would be separated from the "public" state, free from intrusion, unless necessary. This notion of *parens patriae*,[1] embedded a dichotomy between state care and family care.

This assemblage of governing mentalities were embedded into new theories of education for young children as well as older children. They were used as rationales for intervention by philanthropists, social reformers, police, lawyers, and physicians in the nineteenth century as religious and idealist ways to save the child and family for the betterment of an imagined cosmopolitan, homogeneous, and harmonious American society. In the late nineteenth and early twentieth centuries, in the United States, as well as elsewhere, these philanthropic and religious salvation narratives became secular technologies, described as the rise of the *social* by Deleuze (1979/1997) and detailed in Donzelot (1979/1997), in the case of France. In the United States, the new rationalities about the social art of governing included the concept of *parens patriae*; however, it was the technologies of the school, laws, governmental advice and restrictions, and the newly authorized experts in emerging social science disciplines, psychology, education, sociology, social welfare, politics, law, economics, and pediatrics that became the methods of intervention into *private* families, when abnormality was detected. The new social sciences could define and predict generalizable truths along with scientific strategies, to observe populations, group them, survey them, and regulate them. The use of statistics grew and shifted in scope. Whereas statistics had been used in the nineteenth century in the census to define populations, populational reasoning, the science of risk, and probabilistic reasoning came to be used to focus reforms on social problems, unhealthy situations, and dangerous, uncontainable populations.

New institutions and professions of experts grew. The U.S. *Children's* Bureau (begun in 1916 by progressive social reformer and feminist Julia Lathrop) organized massive mailings of parenting pamphlets and federal guidelines for doctors, lawyers, educators, social workers, and parents defining normal parenting, normal childhoods, and the development of moral, healthy, normal (assimilated) citizens. At a time of tremendous population shifts into cities, these changes, including the growth of kindergartens as a way to Americanize children (and their parents), served as the basis for changes in university programs for middle-class women (future mothers) as well as for the poorest and most "abnormal" families, whether defined by widowhood, single parenthood, poverty, language, ethnic background, or race. New laws and regulations were designed to regulate the welfare of these populations; the technologies of governing embedded the earlier dichotomies (private versus public, family versus state, civil society versus the state, individualism and responsible autonomous free choices versus the collective) as ways to regulate subjectivities about others, as well as to govern individual/family behavior.

Governing and Government

The art of government . . . is concerned with . . . how to introduce economy, that is the correct manner of managing individuals, goods and wealth within the family, . . . how to introduce this meticulous attention of the father towards his family, into the management of the state.

(Foucault, cited by Rabinow, 1984, p. 15)

The technologies that came together then included new rationales about where the *body* of the young child should be, how he/she should be scrutinized, observed, judged, and regulated. Disciplinary technologies were not repressive punishments but strategies for dispersing a gaze that would subjectify, police, and differentiate. To illustrate these shifts and ruptures, I turn to a more detailed discussion of Froebel, Dewey, and Addams as illustrations of broader patterns.

Froebel and His German– and English-American Apostles

Catching Children Early: Natural Stages of Development for Two- to Seven-Year-Olds

Froebel's kindergarten, for two- through seven-year-old children, was developed as a result of his work at the University of Jena with German transcendental idealist philosophers Kant, Fichte, and Schelling, his German Pietist religious beliefs, and his work, in Switzerland, with Pestalozzi's educational program that drew directly from a Rousseauian philosophy of children learning naturally through empirical relationships with objects. The political, philosophical, and socioeconomic revolutionary elements combined in the separated Germany resulting in a radical or revolutionary movement by some to unify separated principalities together with one constitution and one ruler in 1848.

The assemblage of different discourses surrounding the beginning of the kindergarten included the different philosophies, gendered discourses about childhood, family, and spirituality, as well as shifts in politics, the growth of sovereign nation-states, and the economy. Industrialization and science, as well as factory work outside the home changed the nature and locations of "family" work and life. The rise of an evolutionary notion of science in the early nineteenth century, a form of cultural recapitulation theory, was also related to imperialism, colonization, and slavery that embedded senses of racial superiority into ideas of childhood and national development.

Froebel's pedagogical ideas focused on the natural instincts and stage of early childhood as a special period of development and education for future citizens. His ideas combined a sense of knowledge that integrated the philosophic principles of an inner and unified knowledge with new scientific ideas of objective sensory, empirical experiences, and the natural. His theory of play was evolutionary; play was seen as a stage of primitive development in which young children, more primitive animal species, and more primitive people around the world were engaged. These came together with the discourses of being part of a greater whole—a unified Germany and a spiritual world.

Evolutionary theories of social/cultural development of the species, as well as races, circulated, reinforcing the social superiority of civilized and cultured societies (sometimes referred to as races). These were embedded in early social biological discourses related to *different* ethnic, religious, and colonized groups. Tröhler (2003) suggests this was certainly true in German discourses related to the superiority of who was cultured and who or what was not.

Froebel's principles of "natural" education were embodied by his structured *gifts* and *occupations* that were to cultivate young children's natural instincts toward *self-directed* and internalized (autonomous/rational) activity, as well as to develop the young child, like a well watered seed, into a future citizen. This "natural" child was part of the universe, of mankind, and of God. Froebel's *Pedagogics of the kindergarten* (1861/1897) emphasized child development, and the need to recognize young children's education as a special stage of learning, in need of its own forms of pedagogy. The following illustrates these points:

> "Come, let us live with our children," becomes, when manifested in action, an institution for fostering family life and for the cultivation of the life of the nation, and of mankind, through fostering the impulse to activity, investigation, and culture in man; an institution for self-instruction, self-education, and self-cultivation of mankind, as well as for all—therefore for individual cultivation of the same through play, creative self-activity, and spontaneous self-instruction . . .
>
> Man, as a child, resembles the flower on the plant, the blossom on the tree; as those are in relation to the tree, so is the child in relation to humanity—a young bud, a fresh blossom; and as such, it bears, includes, and proclaims the ceaseless reappearance of new human life . . .
>
> But man is a created being, and, as such, is at the same time a part and a whole (therefore, a part-whole) . . . he is, as a creation, a part of the universe; but, on the other side, he is also a whole, since—just because he is a creature—the nature of his Creator . . . lives in him. (Froebel, 1861/1897, 6–8)

In the *Education of Man* (Froebel, 1826/1887), Froebel illustrates clearly the influence of science, and social evolutionary theories, including the theory of the recapitulation of the races, suggesting young children and their play were similar to primitive species (and peoples).

> Man, humanity, in many, . . . should, therefore, be looked upon not as perfectly developed, not as fixed and stationary, but as steadily and progressively growing, in a state of ever living development, ever ascending from one stage of culture to another . . . Indeed each successive individual being . . . must pass through all preceding phases of human development and culture (pp. 17–18).

Traveling to America

Froebelian ideas were embedded in the early American kindergarten programs by Froebel's German American disciples and by other European Americans who took up the idea of the new idealist program for young children. The young child's education was to be natural, almost biological, and also self-determining, as well as self-directed. Concurrently, it was structured so that teachers and mothers could *perform* the program appropriately, in the way Froebel's followers intended, through scripted activities, moral lessons in songs, finger plays, stories, gifts and occupations, and "free" natural play. In books translated from German and in later kindergarten teacher trainings, directions were provided as to how and what mothers or kindergarten teachers were to teach, stories or songs to use, and how gifts and occupations should encourage child-directed play (Froebel, 1861/1897).

"*Gliedganzes* in Froebel's meaning signifies that man is a whole or self-determining being and at the same time a member of a social whole." Tröhler (2003) suggests this was an important part of German philosophy relating to the juxtaposition of "empiry and *geist*, plurality and unity" with German philosophy and politics emphasizing unity and culture over the empirical and plural. However, this was also a pedagogical and philosophical difference that appeared in German Froebelian philosophy as it traveled to America and was *translated* by American Froebelians in the late nineteenth century. The German ideas encountered American pragmatic educational theories that emphasized the individual in social experimentation, the scientist, the empirical over the humanities instead of a unity and spirituality of culture.

The American Froebelian program used disciplinary technologies to fabricate which types of education were "best," which families and children were well educated, and which families, mothers, and children required interventions through parent education or the *kindergarten* remediate deficiencies of the home. In the United States, the kindergarten was translated and hybridized into the local cultural imaginaries of what education should be for the imagined citizen and American nation of the future.

Disciplining the Body While Saving the Child, Family, and Nation

The pedagogical discourses that embedded modern German philosophers' thinking into the new American kindergarten program for young children, were integrated with an America that was shifting and integrating religious salvation beliefs, idealist and transcendentalist philosophies, and a pragmatic, empirical, and evolutionary science. Industrialization was built on scientific experimentation, and notions of making progress through science.

The late nineteenth-century United States was a country searching for modern education that would discipline and govern citizens for their own welfare as well as for the nation's. Catching children early through a new stage of preschooling became more popular by the 1870s because of increasing heterogeneity in the country, in general, and especially in growing urban areas. During this time, an increasingly visible presence of African American and other immigrant children in urban schools throughout the country created a sensation of "danger" to many members of early settler groups.

The Americanized Froebelian kindergarten was recognized increasingly as an intervention into family life (on behalf of the nation) that was a more *natural* place for young American children to play and learn and to become disciplined in moral habits, cleanliness, and manual skills and occupations. As the kindergartens grew, they moved more and more out of private settings and into the public schools as the first preprimary stage at which young (4–5 year old) children could be assimilated and their active bodies, languages, and habits could be tamed or civilized.

Policing and Corralling Children

By the turn of the century, American children were placed spatially together in supervised settings such as kindergartens, schools, and playgrounds; this removed children from streets and factories, and the kindergarten controlled young children through early English language teaching as well as Americanization—habit training, disciplining minds, morals, and, in the kindergarten, through manual dexterity training in the fine motor oriented gifts and occupations that Froebelians increasingly used as the focus of curriculum. A brief excerpt citing Felix Adler's ideas about the kindergarten in the late 1890s in New York illustrates these points:

> A pauper class is beginning to grow up among us, incapable of permanently lifting themselves to better conditions by their own exertions . . . only rendered the more dangerous and furious by the sense of equality with all others, with which our political institutions have inspired them . . . of all these possible measures of prevention, a suitable, a sensible system of education is assuredly the most promising. (Adler, "Free Kindergarten and Workingman's School" cited in Bloch, 1987, p. 39)

The American kindergarten was used to fabricate subjectivities and desires to *be* American, as well as to be certain kinds of future workers/citizens. It was constructed by Froebel's disciples as a modern, idealist, and transformational program for young children, worthy of public as well as private (philanthropic) funding. Its pedagogies embedded discourses that appeared to be inclusive of all but included different reasonings intended for populations conceived of as unequal.

Technologies of Science: Empirical Observations and
Assessments as Strategies of Differentiation and Surveillance

Technologies of natural science that had been used to observe cross-species development, promulgated by Darwin (1859), reinforced the natural stage of evolutionary development, primitive play, and instinctual unreasoned behavior that characterized *all* young children. Recapitulation theory was used to reinforce the natural superiority of highly educated, cultivated European and English-born citizens over "others," new to cities, and to the United States. In the United States, social Darwinism and science intertwined to imagine developmental stages scientific and objective and accepted universal truth defining children and childhood.

The Social Administration of Freedom: John Dewey and Jane Addams

The Rise of the Social: Governing Children's Free Play and Activities

The new institutions and technologies that developed in the late nineteenth century were considered secular approaches to saving the young child and his/her family; a rupture in the rationalities and technologies used to discipline and civilize souls and subjectivities occurred. The assemblage of secular discursive languages and practices that rose at the turn of the century was related to the governing of the welfare of the nation and its populations. New scientific fields and techniques allowed for a populational and statistical reasoning that could define groups more precisely and through that appropriate places, times, and ways to intervene.

Stages of Distinct and Universal Development

In the early twentieth century, *preschool age* children as well as *adolescent* children were separated from those of *elementary age* as different populational groups. The groups were defined by different developmental characteristics, described by scientific theories, observations, questionnaires, and tests, categorized as normally or abnormally developed toward maturity, along a linear, progressive pathway. Maturity was defined by age and various attributes, skills, and conduct (see Bloch, 1987 on G. Stanley Hall, and the growth of scientific observations during the first quarter of the twentieth century).

The Normal Family

European-American autonomous, responsible, and economically self-sufficient (male) adults were the ideal citizen, and ideal head of the American family; normal wives and mothers were still to be at home caring for the future citizens, their children. America also fabricated a normal family as one that could take care of itself; dependence upon the state, or even from philanthropic charity, was perceived as a mark of an abnormal family or individual. Therefore, widows, or single parents/mothers, as examples, were constructed as abnormal families, even if they were independent and economically responsible. Early social welfare policies focused on provisions for women to take care of children at home; day nurseries were developed as support for women/families where this could not happen. Abnormal families and children were also marked by ethnicity or language; American Indians, Chinese immigrants, African Americans, Irish, and Eastern European immigrants were homogenized as abnormal and in need of different forms of state intervention.

Governing the Freedom of the Child

In *Governing the soul* (1989), Rose shows how the early disciplinary technologies included observations, testing, interviews, and questionnaires. These were used to establish universal guidelines for development within and across stages, as well as normal and abnormal development. Scientific observations in University preschool laboratories, objective science on selected groups of children could help decipher what scientifically knowledgeable parents should do and be like; the compilation of statistics of what the average child did at 1, 2, 3, 4, or 5 years of age became part of a universal guideline of what all children should be doing. While these guidelines became established by the 1940s, they were being developed during the first quarter of the twentieth century under the tutelage of G. Stanley Hall, John Dewey, Patty Smith Hill, and others working in University based preschool settings. The statistical tests and observations of middle-class Euro-American children resulted in generalizable laws of normal development, which came to be interpreted as "universal"; these norms and expectations for universal development are still largely what are used today to guide parents, in teacher education, and in programs focused on developmentally appropriate practices and good child development.

But the early twentieth century progressive discourses embodied a rupture with the past ideas of the Froebelian kindergarten and yet enhanced the importance of science as a way to determine truth. Child development was to flow through stages from immaturity to maturity, from undeveloped to developed, as in the Froebelian programs, but now this was scientifically defined truth, based upon empirical observations.

John Dewey's work at the University of Chicago, as well as Jane Addams' social work at the Hull Street Settlement House in Chicago represented some of the ruptures from discourses of the Froebelian program. Both Dewey and Addams focused on the importance of education being pragmatic, connected with life, not abstract, more secular, scientific, and based on *real* experiences, and objective empirical data. Whereas Dewey promoted these ideas in his Laboratory School, Addams promoted these discourses in her work with poor and immigrant families, women, and children at the Hull Street Settlement House, where a kindergarten and day nursery were part of a more general institution to serve families (Addams, 1910/1990.)

Governing the future citizens' subjectivities such that they would take their roles in this new society for granted, not question their place in capitalist society—in the factory, in the schools, and in the kindergartens—was crucial. Because it was important to see oneself as free to choose, to act, and to participate in the new American liberal democratic and capitalist society, it was also important to devise new technologies for the social administration of freedom, conduct, and subjectivities of what was normal, and good—for children as well as their families, in schools, and for teachers.

Citizenship for a Progressive and Democratic Nation: The Scientific Child
John Dewey's secular, pragmatic, and empirical approach to solving social problems included a national imaginary of a more inclusive democracy that would serve everyone equally and be less divided along class and other social lines. In his writing, he opposed both the notion of superior culture, in principle, as well as the notion of different education for different children, or a tracking of different children for different roles and different class-based jobs. His imagination of democracy and education was one that would include everyone in social decision-making, as a miniature social community. Children's participation in civil society, even from the youngest ages, was to inculcate a sense of citizenship. He hoped that children would

take their own experiences and interests as the base for activities and that with the guidance of teachers education and learning would emerge from children's active engagement of studies of social life. He also wanted children to cooperatively solve social and other problems, appropriate to their developmental level. All of these ideas were embedded in the Chicago Laboratory School for the children of faculty and staff at the University of Chicago.

From this position, Dewey believed in the possibility of laws of universal child development, but this was not enough. The child's character must be shaped by offering the right types of experiences to create concern for society and community (Bloch & Kennedy, 2001). He believed the Froebelian programs were "cumbrous and far-fetched, giving abstract philosophical reasons for matters that may now receive a simple, everyday formulation" (Dewey, 1915/1990, pp. 121–122). In his book, *The school and the child*, Dewey highlighted the growing importance of science in his own interpretations of what and how a child learns or is educated:

> it is hardly likely that Froebel himself would contend that in his interpretation of . . . games he did more than take advantage of the best psychological and philosophical insight available to him at the time; and we may suppose that he would have been the first to welcome the growth of a better and more extensive psychology (whether general, experimental, or as child study). (Dewey, 1915/1990, p. 121)

In *Democracy and education* (1916), Dewey's enthusiasm for scientific inquiry is in order to make a particular type of future citizen. By the mid-1930s, he was even more convinced that social science and empirical experimentation would help to solve the social problems of poverty, particularly through individual and cooperative problem-solving. He was also concerned that social sciences should have been able to provide more solutions already for the many political, social, and economic problems faced in the America of the early 1930s.

Dewey acknowledged that *all* children and families were not the same in the early twentieth century, but he aspired toward a liberal rationality of equality of opportunity, within real-life experiences and education that was functional and pragmatic, theoretically, based on both children's interests, and teachers' ideas about democracy and education. Dewey, aspired to use young children's developing curiosity, their play, their imagination, and their interests and experiences to form new types of citizens; their aspiration was to erase differences, rather than to reinforce them. The approach was to govern young children to become *self-governing*, to solve problems themselves, to take responsibility for their own actions, to gather empirical observations about the real world, and to use science to hypothesize about different ways to solve mundane and more serious problems.

The first aim of the kindergartner (the teacher) should be to form a social atmosphere, and make her kindergarten as much a home as possible. In order to do this there must be freedom and as few rules as the surroundings will permit, thus throwing each child upon his own responsibility and allowing him opportunity for expressing individual traits of character. This gives the child an opportunity for forming laws and rules of his own (Scates, 1900, p. 120).

Universal Development and Common Experiences:
Governing Freedom

Dewey's ideas are often interpreted in terms of freedom, the liberty of the individual child, cooperative problem-solving, and a miniature community that was to imitate and teach children how to participate in a democratic society, that was, he admitted, a work-in-progress, still unequal, particularly in terms of economics and social standing (education, culture). In the earliest years, this miniature society was based on real (not abstract) open-ended imaginative play, construction work with large blocks, and reenactment of cooking and other activities that children experienced at home and neighborhood. These ideas emphasized learning by doing; learning to be self-governing and, as a group, to govern each other (see Dewey, 1902, 1915). While critiques of Dewey's notion of child-centeredness, the class, race and gendered discourses reembedded in many activities exist (e.g., see Walkerdine, 1984, for one example), the rationalities of child-centeredness, and freedom of choice remain inscribed, without critique, in current documents describing "best practices"and high quality early childhood programs.

Jane Addams and the Hull House

Jane Addams, one of the most famous of the early educated woman social reformers, was a colleague of Dewey's in Chicago. While Dewey taught philosophy and education at the University of Chicago, Addams, as a college educated woman, had aspirations to do more than "simply" get married and do philanthropic work. She hoped to use her education as a model for other elite women, and to help others through building a community filled with activities, and possibilities for saving the urban immigrant poor from poverty, child labor, abusive labor practices, and providing minimal wages. Her educational program was directed at children, their mothers, their fathers, and their families, at immigrants surrounding Chicago. It was aimed at the elite, governing the way they would think about social welfare and reform efforts based on new images of a new America, less based on class, and individual privilege, more social, aimed at a hierarchically organized "collective." She surveyed workers' hardships and tried to use secular technologies to remedy many of those through social legislation, new institutions, and a new imaginary of the social responsibilities educated women should have (see Addams, 1910/1990).

Parens Partite in action

Both Dewey and Addams felt that it was appropriate to intervene through education, and other means, to provide opportunities that immigrant and other impoverished families in urban Chicago could not provide. Instilling new skills and attitudes in children, when young, was one of the best ways to intervene into families and children's lives, and subjective identities. Both Dewey and Addams thought it was appropriate for modern secular society to develop new institutions and new interventions to help develop self-governing skills and attitudes, but the direction of self-governance, of course, should be to take on appropriate cosmopolitan, "American" values.

Addams especially aimed at education for reforming society. She believed science could be used for the redemption of secular souls. She aimed to help the poor and to reconstruct "experiences" and interests of immigrant children and their families in order to reproduce American (homogeneous) culture. In her work, we see a universalized notion of motherhood and childhood for a new society. Her direct work focused on reforms to limit child labor, exploitation of factory workers, especially women, in terms of the long hours of work required, increasing wages for work as well as the quality of conditions at work. Hull House, through their myriad activities and opportunities provided shelter for homeless and destitute, as well as for abused women and their families. A day nursery to support women's labor in factories was opened that also allowed older children to go to school; in addition, a kindergarten was begun to experiment with new progressive educational ideas, also helping to catch children early to reconstitute experience for new, and, perhaps, better American citizenship (Addams, 1910/1990).

Educating all Men (and Children): Governing Freedom from Afar
Addams' work, was to govern the new citizen from afar. Education was the means by which immigrants and their children might hope to achieve something better than exploitative factory work, poverty, hunger, or religious/political oppression. As Addams (1902/2002) writes:

> As democracy modifies our conception of life, it constantly raises the value and function of each member of the community, however humble he may be . . . We are gradually requiring of the educator that he shall free the powers of each man and connect him with the rest of life . . . we have become convinced that the social order cannot afford to get along without his special contribution . . . as we throw down unnatural divisions . . . in the spirit of those to whom social equality has become a necessity for further social development (p. 80).

Socially Administering Freedom as Part of Modernity

Attention focused on the scientific, progressive, democratic citizen who would have equal chances to become part of the homogeneous cosmopolitan prosperous culture that a liberal free-market capitalist society appeared to offer as part of its imaginary. The contradictions were in governing the self-active, problem-solving maturing and mature civilized individual for a cosmopolitan nation that also was built on stratification. Self-governing individuals had to learn to cooperate with each other and to take the liberal hope for equal opportunity through education and self-improvement, becoming normal in America as the way to find opportunity. Social and economic conflicts were to be minimized; governing freedom and the general welfare of the population through early and later education was an important technology for governing harmony itself. There was little space to question the exclusions built into the new technologies and rationalities of individual promise and progress.

While there were differences in approaches to education, modern education from Froebelian times through the early twentieth century embodied the reasoning of the civilized rational autonomous self-governing man, a developing child passing through biological and cultural stages from less to more civilized citizenship. Reasoning reinforced the idea that liberal individualism would serve capitalism and democracy.

The Growth of Global Uncertainty in Late Modern
and Post-Modern Moments

Today, late modern, post-Fordist capitalist democracies still appear to look for a standard, universal well-developing child as future citizen. However, reasoning about the child now includes an orientation toward greater flexibility, entrepreneurial abilities, and the multicultural child, understanding, knowledgeable about, and tolerant of difference. Today's educational reforms also express new cultural anxieties, new uncertainties, and new tensions between an imagined, romanticized past that looks backward to the imaginary of a homogeneous, and harmonious society, while we also imagine education as preparation of children for uncertain, entrepreneurial, competitive, pluralistic, and globalized futures. As in the past, however, the standard, developing child embodied in current educational forms, such as *No Child Left Behind*, are to catch "different," abnormal children (and their parents) as early as possible to prepare them at a *proficient* level (in literacy as well as "morality") to participate in the imagined nation of the future.

The technologies today are both the same (testing, observation) and different than in the past. We have moved from disciplinary societies and a disciplinary notion of power/knowledge to a period when both modernist disciplinary strategies as well as more encompassing and penetrating technologies that regulate us from our "neurons to neighborhoods" are increasing. New medications are used to control the young child's body; home and school educational pedagogies representing "developmentally best practices" govern parents, teachers, and children's desires, bodies, conduct, and souls (Foucault, 1977, 1988; Rose, 1999).

The Rhizome as a Metaphor for Opening Up Childhood and Education Rather Than a Closing Down.

If we imagine a more postmodern moment, we can examine other differences in discourses that are now circulating, controlling, as well as opening new spaces for action and thought. Deleuze (1990) suggests that with increased globalization of communication, cultures, economies, that what we are experiencing are "societies of control" that involve myriad new methods of control, as well as places for eruptions, and rhizomatic, unpredictable growths, and "becomings" (Deleuze & Guattari, 1987). Manuel Castells (2000, 2004) suggests a move toward spatial flows and network societies (Castells, 2000, 2004), in contrast to the concept of the post-Fordist information or knowledge societies in which we engage in lifelong learning, and educational self-governing along the developmental lifespan. Do current discourses of a modern standard and universal best practice in schooling for young children (e.g., Bredekamp & Copple, 1997, *No Child Left Behind* reforms in the United States, voucher systems to provide choice, testing and assessment to construct normal plural but homogeneous citizens) reflect these new openings for conduct, identities, control, and action?

Catching the Young Child and His Family Early

Today each one of us must be responsible for ourself and one another. We must be self-governing and help in the governing of each other. New welfare policies in the United States promote responsibility and autonomy from the state by everyone; there are no more guarantees of state support, and low wages, no benefits, and responsibility for one self and family have become increasingly normal for a larger part of society. As globalizations of economies, outsourcing, and cultural and physical border crossings are enhanced, the young child and his parents must be standard, yet flexible and adaptive, competitive and collaborative, harmonious, able to understand difference, without being conflictual or resistant. Governing from afar, through steering mechanisms such as standards and testing in early education as well as in elementary and secondary education, mimics the circulation of governing technologies in the media, in medicine, in the respatialization and corralling of bodies and minds. Entrepreneurial bodies must fend for themselves in an increasingly competitive globalized world, that also requires some forms of interdependence, while at the same time the sovereign nation-state imagines itself capable of constructing national citizenship in an ever globalized world. In this new world, uncertainty and new openings ask for a privileging of heterogeneity, plurality, and acceptance of complexities of identity and conduct.

The child must be part of a cosmopolitan world; children and their parents must be cosmopolitan, entrepreneurial, and flexible participants in the global economy, political, and cultural context. They are privatized members of a nation that is now part of global society. They are governed to act privately, to choose well in order to be successful for self, nation, region, and world. Teachers are asked to teach children to be members of a globalized world and at the same time to compete well for their own self, family, and nation. The discourses of individual entrepreneurial activity, free enterprise, choice, and competition signal an enhancement of a neo-liberal rationality, that is at the same time universal for everyone and, while seeming to be inclusionary, is exclusionary of diverse cultural ideas, identities, and actions.

The discursive emphasis on hard scientific evidence that can guide educators as well as parents as to how to govern ourselves and our children as well as future citizens has continued into the twenty-first century and perhaps been amplified whereby *rigorous scientific educational evidence* is required of scientists to determine truth and knowledge, as well as who is authorized to speak. At the same time, it should be clear to all that there are multiple truths and that *rigorous scientific evidence* is used only when it serves broader economic, political purposes in the name of welfare for "all."

Universal best educational practices that might get us from point *a* (the young uneducated child) to point *b* (the well-educated developed child—future citizen), embody new disciplinary technologies, reminiscent of the early social efficiency, or manual dexterity movements of the late nineteenth century. Standards and pedagogies are touted as being for all, while, even in early childhood education, current pedagogies are organized differently for those who are poor, different, "at risk," or dangerous, compared to those available privately for children of the middle and elite classes. Governing difference and danger, even when it is of small bodies, is different from the governmentalities of those constructed as "normal."

The Rhizomatic Space of Discursive
Childhood/Citizen, and "Standards for All"

While current standards and reforms suggest that the good future citizen/child will be *proficient* or at least average in terms of certain basic academic and social skills (literacy/math/science, acceptable social behavior), others see a need for more contingent, culturally relevant, and relative education and recognition of complex, contingent identities, other ways to think of high standards for all. Deleuze and Guattari's (1987) notion of the rhizome is used as a metaphor to open new spaces for different conceptions of childhood, pedagogies, multiplicities of identities, conduct, education and care for self and other. Here at the end of this chapter, I emphasize the importance of heterogeneity of childhoods, families, nations, gendered, cultural, geographical, pedagogical spaces over the push for scientifically testable/assessable *universal laws of development* or the *standard, normal child* or *childhood.*

The notion of a rhizome points us toward openings of new possibilities, the blurred, multiple borders, the interstitial spaces, and the multiplicities of identities rather than the dichotomized identities (black versus white, oppressed versus oppressor; male versus female; public versus private; child versus adult; developed versus undeveloped) that often govern thinking and conduct. It pushes toward a need to distance ourselves from a sense of predetermined and fixed ideas of identity, of "the child" or normal development, or one truth about normal to open spaces where nothing is determined, spaces are open (Deleuze & Guattari, 1987). It is with this in mind that I question the rationalities about "good" early childhood education that come to us from the late nineteenth and twentieth centuries, and that appear to remain, in various forms with us still in the early twenty-first century. I believe we should privilege the ruptures and new spaces in thinking and conduct to open different points for conduct, belief, and practice.

Notes

* Thank you to my co-editors and others from the Reconceptualizing Early Childhood Class, Fall, 2005 at the University of Wisconsin-Madison for their comments, help, and discussion. Omissions and faults are my own independent responsibility.
1. According to one recent legal source, "it is Latin for 'parent of his country'. . . (and is) used when the government acts on behalf of a child or mentally ill person. (It) refers to the 'state' as the guardian of minors and incompetent people" (http://www.lectlaw.com/def2/p004.htm). Note: *Child, minor, mentally ill person,* and *incompetent people* are taken for granted here.

References

Addams, J. (1990). *Twenty years at Hull House*. Champagne: University of Illinois Press. (Original work published 1910)

Addams, J. (2002). *Democracy and social ethics*. Champagne: University of Illinois Press. (Original work published 1902) Anderson, B. (1991). *Imagined communities: Reflections on the origins and spread of nationalism* (rev. ed). London: Verso. (Original published 1983)

Bhabha, H. K. (1990). *Nation and narration*. New York: Routledge.

Bhabha, H. K. (1994). *The location of culture*. New York: Routledge.

Bloch, M. N. (1987). Becoming scientific and professional: Historical perspectives on the aims and effects of early education and child care. In T. S. Popkewitz (Ed.), *The formation of school subjects: The struggle for an American institution* (pp. 25–62). Philadelphia, PA: Falmer.

Bloch, M. N. (2003). Global/local analyses of the construction of "family-child welfare." In M. N. Bloch, K. Holmlund, I. Moqvist & T. S. Popkewitz (Eds.), *Governing children, families and education: Restructuring the welfare state* (pp. 195–230). New York: Palgrave.

Bloch, M. N., & Kennedy, D. (October, 2001). *The concept of welfare, child care and early education for children and their families: Replaying the past with different voices but the sametune*. Paper presented at the meeting of the 10th Reconceptualizing Early Childhood Education Conference, New York.

Bloch, M. N., & Popkewitz, T. S. (2000). Constructing the child, parent, and teacher: Discourses on development. In L. D. Soto (Ed.), *The politics of early childhood education*. (pp. 7–32). New York: Peter Lang.

Bloch, M. N., Popkewitz, T. S., Holmlund, K., & Moqvist, I. (2003). Global and local patterns of governing the child, family, their care, and education: An introduction. In M. N. Bloch, K. Holmlund, I. Moqvist, & T. S. Popkewitz (Eds.), *Governing children, families and education: Restructuring the welfare state*. (pp. 3–31). New York: Palgrave.

Bredekamp, S., & Copple, C. (1997). *Developmentally appropriate practice in early childhood programs*. Washington, DC: National Association for the Education of Young Children.

Castells, M. (2000). *The rise of the network society, The information age: Economy, society and culture* (Vol. 1) Cambridge, MA: Blackwell. (Original work published 1996)

Castells, M. (2004). *The power of identity, the information age: Economy, society and culture* (Vol. 2) Cambridge, MA: Blackwell. (Original work published 1997)

Chakrabarty, D. (2000). *Provincializing Europe: Post-colonial thought and historical difference.* Princeton, NJ: Princeton University Press.

Darwin, C. (1859). *Origin of the species.* London: Charles Randolph.

Deleuze, G. (1990). Postscript on control societies. In *Negotiations* (pp. 177–182). New York: Columbia University Press.

Deleuze, G. (1997). Forward: The rise of the social. In J. Donzelot, *The policing of families* (pp. ix–xviii). New York: Pantheon Press. (Original work published 1979)

Deleuze, G., & Guattari, F. (1987). *A thousand plateaus: Capitalism and schizophrenia.* Minneapolis: University of Minnesota Press.

Dewey, J. (1902). *The child and the curriculum.* Chicago, IL: The University of Chicago Press.

Dewey, J. (1915). Froebel's educational principles. In *The School and society.* Chicago, IL: University of Chicago Press, 11–27.

Dewey, J. (1966). *Democracy and education. An introduction to the philosophy of education.* New York: Free Press. (Original work published 1916)

Dewey, J. (1990). The School and society. Chicago: University of Chicago Press (original work published in 1915)

Donzelot, J. (1997). *Policing the family.* Minneapolis: University of Minnesota Press. (Original work published 1979)

Foucault, M. (1977). *Discipline and punish.* New York: Vintage Books.

Foucault, M. (1980). *Power/knowledge: Selected interviews and other writings, 1972–1977* (C. Gordon, Ed.). New York: Random House. Foucault, M. (1988). Technologies of the self. In L. Martin, H. Gutman & P. Hutton (Eds.), *Technologies of the self* (pp. 16–49). London: Tavistock.

Foucault, M. (1991). Governmentality. In G. Burchell, C., Gordon & P. Miller, P. (Eds.), *The Foucault effect: Studies in governmentality.* (pp. 97–104). Chicago, IL: The University of Chicago Press.

Froebel, F. (1887). *The education of man.* New York: D. Appleton and Company. (Original work published 1826)

Froebel, F. (1897). *Pedagogics of the kindergarten: Or, his ideas concerning the play and playthings of young children* (J. Jarvis, Trans.). London: Edward Arnold. (Original work published 1861) Retrieved December 10, 2005, from http://wordsworth.roehampton.ac.uk/digital/froarc/froped/ind.asp

Locke, J. (1970). *Two treatises of government* (2nd ed) (Peter Laslet, Ed). Cambridge: Cambridge University Press. (Original work published 1689)

Locke, J. (1999). Some thoughts concerning education. In J. W. Yolton & J. S. Yolton (Eds.), *Clarenden Edition of the Works of John Locke.* Oxford: Oxford University Press. (Original work published 1693)

O'Malley, P. (1998). Indigenous governance. In M. Dean & B. Hindess (Eds.), *Governing Australia* (156–172). Cambridge: Cambridge University Press. Popkewitz, T. S. (2000). *Educational knowledge: Changing relationships between the state, civil society, and the educational community.* New York: State University of New York Press.

Popkewitz, T. S., Franklin, B. M., & Pereyra, M. A. (2001). *Cultural history and education: Critical essays on knowledge and schooling.* New York: Routledge.

Popkewitz, T. S., & Bloch, M. N. (2001). Administering freedom: A history of present rescuing the parent to rescue the child for society. In K, Hultqvist and G. Dahlberg (Eds.), *Governing the child in the new millennium* (85–118). London: Routledge.

Rabinow, P. (1984). *The Foucault reader.* New York: Pantheon Press.

Rousseau, J. J. (1968). *The social contract.* (M. Cranston, Trans.). NewYork: Penguin Books. (Original work published 1762) Rousseau, J. J. (1979). *Emile, or on education* (Trans. with an introduction and notes by A. Bloom). New York: Basic Books. (Original work published 1762.)

Rose, N. (1989). *Governing the soul.* Cambridge: Cambridge University Press.

Rose, N. (1999). *Powers of freedom: Reframing political thought.* Cambridge: Cambridge University Press.

Said, E. (1978). *Orientalism.* New York: Vintage Books.

Scates (1900). School Reports: The sub primary kindergarten department. *Elementary school record.* June, 129–142.

Tröhler, D. (2003). The discourse of German *Geisteswissenschafteliche Padagogik*—A contextual reconstruction. *Paedagogica Historica, 39* (6), 759–778.

Walkerdine, V. (1984). Developmental psychology and the child-centered pedagogy: The insertion of Piaget into early education. In J. Henriques, W. Hollway, C. Udwin, C. Vena, & V.Walkerdine (Eds.), *Changing the subject: Psychology, social regulation, and subjectivity.* (pp. 153–177). New York: Methuen Press.

Wollons, R. (2002). *Kindergarten and cultures: The global diffusion of an idea.* New Haven, CT: Yale University Press.

Young, R. J. C. (1995). Culture and the history of difference. In *Colonial Desire: Hybridity, Culture and Race.* (pp. 29–55). London: Routledge.

CHAPTER 2

Configuring the Jewish Child: Intersections of Pedagogy and Cultural Identity

DEVORAH KENNEDY

Introduction

Child care quality is a central feature of the Bush administration's reform initiative in early childhood education.[1] In this chapter I discuss assumptions embedded in discourses of child care quality and power effects of current discourses in relationship to cultural difference. I approach the concept of "child care quality" first as a product of historically constituted reasoning inscribed with assumptions and comparative differentiations between "normal" and "non-normal" populations and individuals. As such quality child care becomes a governing concept, delineating parameters through which we guide our own behavior and judge that of others. In particular, I approach "quality" child care and the notions of "diversity" and "inclusion" produced through those discourses, as productive of parameters defining "normal" childhood and acceptable "cultural difference." I look specifically at production of configurations of normal Jewish childhood within the American Jewish community.

In the first section of the chapter I discuss my analytical approach. In brief, I approach "quality child care" as discursive practices constituted within intersecting trajectories of historically constituted reasoning. The fluid and contingent interactions of these trajectories produce conditions making possible the production of new knowledges, differentiations between populations and individuals, and broad understandings of the world. In the second section, I discuss historically constituted knowledges as conditions of possibility for current concepts of "quality child care." The discussion focuses on differentiations assumed within the reasoning, making and drawing linkages between those differentiations and configurations of Jewishness. In the final section, I discuss current discourses of quality child care as governing concepts that produce parameters defining "normal" childhood. In particular, I look at diversity and inclusion as they are incorporated within the discourses and examine effects of this reasoning on notions of Jewishness and Jewish difference.

Analytical Approaches: Modern Systems of Reasoning

In this section, I outline theoretical and philosophical concepts underlying the chapter. My analytical framework combines Foucault's approach to historical analysis of current problems with Bhabha's and Chakrabarty's explorations of encounters between "Western" and "other" modern systems of reason. I draw on Foucault to address historical shifts and ruptures in European reasoning conditioning our understandings of ourselves as "modern" (see for example, Foucault 1966/1973, 1975/1995). Additionally, Foucault's work clarifies rules governing the production of truth and the power effects of knowledge (i.e. Foucault, 1980, 2003). I also look to postcolonial/subaltern studies writers such as Bhabha (1994) and Chakrabarty (2000, 2002) to address issues of cultural difference within modern systems of reasoning. These theorists address the complex relationship between "modern," "Western" reasoning and systems of reasoning of colonized "other" populations. The focus on power relationships between and within systems of reasoning enables me to analyze complex processes. Rather than focusing on structural spatial and racial notions of oppression, I explore the reasoning through which these concepts are constituted.[2]

Foucault: Historical Change, Truth, and Power/Knowledge

Throughout the chapter, I draw on Foucault's work to make several points. First, by tracing the emergence of rules and standards for the production of knowledge and truth Foucault (1966/1973) focuses attention on major cultural/intellectual/social shifts that have characterized the history of European ("Western") systems of reasoning. Through these shifts and ruptures conceptualizations of knowledge and truth were reconstituted from forms in which gods, truth, and representation were mutually reducible into systems of reasoning distinguishing universal, objective, reasonable truth and knowledge from particular personal beliefs including culture or religion. Describing changes as ruptures rather than progress makes clear that prior systems of reasoning are not simply superseded by more advanced rational or scientific knowledge. Rather, ruptures or discontinuities indicate change in what we understand as "truth" and the rules for its production. However, existing truths and knowledges do not disappear, they are re-inscribed in different forms within new rules for production of truth. Hence, new knowledges are also inscribed with pre-existing assumptions.

Another crucial concept I draw from Foucault's work is "governmentality" (1979/1991). Governmentality describes relations of power between government, knowledge, and conduct crucial to notions of progress and freedom. Modern notions of progress, freedom, and reasonable conduct emerged through shifts in reasoning that included the following:

1. An idea of governmental power as concerned with responsibility for its population. Governments' decisions concerning care of populations were to be based in expert knowledge about populational and individual needs.

2. Knowledge and truth understood as scientifically discovered rather than divinely revealed. Expert knowledges assert norms about appropriate, desirable behavior for individuals and populations. Normalizations of rational conduct emerge in comparison to, and produce notions of, the conduct of non-normal populations. In comparison to normal populations, "others" are conceptualized as savage, superstitious, emotional, or unreasonable; but, in any case, the populations they describe are often conceptualized as "lacking," "deviant," or even "dangerous."

3. Re-inscription of Protestant "pastoral power" and individual redemption within emerging modern reasoning as a type of self-responsibility (Carrette, 1999). This re-configuration of individual redemption as self-responsibility makes possible modern notions of freedom through production of the independent self-reflective actor whose behavior is guided through reason and knowledge (Popkewitz, 1998). Through these shifts the family emerged as "the privileged instrument for the government of the population" (Foucault, 1979/1991, p. 100). Expert knowledges related to child health, growth, and development established in medicine, and later psychology, produce dispositions, desires, and understandings that guide, and through which we evaluate, individual behavior of family members and educators (see Rose 1990, 1996). This reasoning is evident in current "quality child care" discourses. "Quality" is assumed to be determined in a logical, objective, scientific, universal manner. We assume it can be rationally determined, measured, and standardized, and that individuals will make decisions based on this knowledge. Despite critiques of current notions of child care quality on the basis of affordability, location, and cultural preference,[3] parents' choices allow families and children to be ranked and ordered based upon the quality of their child care arrangements (Kennedy, in press).

Bhabha and Chakrabarty: "Western" Reason and "Others"

Foucault shows the historical contingency of "modern" "Western" reasoning, and the ranking and ordering of populations occurring therein. However, other populations and individuals are not blank slates produced solely within constraints of modern reasoning (Chakrabarty, 2002). For example, Sartre's depiction of the Jew (1948) constitutes Jewish difference primarily through "Western" anti-Semitism.[4] However, Judaism is not an empty other to be signified through Western knowledge (whether modern secular or traditional religious). Rather, different Jewishnesses are constituted through encounters between Judaism as a system of reasoning and the reasoning of dominant populations within which Jewish communities are

46   Kennedy

situated, including "Western" systems of reasoning. Thus, I draw upon the
work of Bhabha (1994) and Chakrabarty (2000, 2002) to examine encounters of modern reasoning about "child care quality" and Judaism as an
"other" reasoning in the production of cultural difference.

Crucial to my analysis of encounters between systems of reasoning is
Chakrabarty's (2000) argument that "modernity" conflates "historical" and
"cultural" time (i.e. "Western" becomes the equivalent of "modern").
"Western" reasoning conceptualizes itself as modern, civilized, and rational
through comparison to "other" systems of reasoning about the world,
ranking and ordering "other" populations and their reasonings along a pre-
sumably universal developmental, progressive time continuum. In effect,
this relegates "other," but contemporaneous, systems of reasoning to
"primitive" or "pre-modern" status. Through such reasoning "other" systems of understanding the world become conceptualized through categories thought to be personal rather than universal including beliefs,
cultures, and religions and are ordered along developmental continua ranging from primitive to rational. "Others" are positioned, and position themselves, along the continua through such comparative and judgmental terms
as fundamental, traditional, progressive, and secular.

Difference, Mimicry, Translation, and Hybridity

To understand processes by which "other" systems of reasoning come to
know themselves through categorizations typical of "modern" "Western"
reasoning I draw on Bhabhas' (1994) analyses of cultural difference. Bhabha
distinguishes between cultural diversity and cultural difference: cultural difference is a productive act, an act of enunciation, whereas diversity as an
epistemological object in which "culture is an object of empirical knowledge" (Bhabha, 1994 p. 34). For example, notions of diversity circulating
in education assume "cultures" have identifiable characteristics. Thus, we
should be able to add knowledge of these characteristics to curricula as
diverse variations of human behavior within parameters of "normal" childhood and universal human development. In effect, this produces parameters for acceptable difference, re-iterating and re-inscribing "Western"
cultural values prevalent in knowledges asserting universal notions of
progress, rationality, and development.

On the other hand, cultural difference refers to processes of cultural
enunciation. An enunciation is a statement that incorporates multiple
domains, in which objects appear and which assigns power relations
(Foucault, 1971/1972, p. 89). Thus, an enunciation, whether about childhood or Jewishness, delineates possibilities and shapes meanings, establishing parameters of inclusion/exclusion and relations of power. Enunciations
of cultural difference involve both assigning comparative status to "others"
and translating "other" systems of reasoning into universal categories.
Discussions of cultural difference necessarily create parameters for including/
excluding and ordering populations and individuals.[5]

When systems of reason encounter each other translation occurs, and hybrid reasonings are produced in two ways. One, dominant reasoning translates "other" reasonings into its own comparative representations and categorizations of differences. This process is evident in my discussion of the conflation of cultural and historical time. Two, the "other" system is changed through translation into categories/rules/standards of the dominant reasoning. For example, Jews have been described as having had unity of gods/knowledge/representation revolving around Israel, Torah, and God (Lipset, 2003). In translating these concepts into "modern" categories of reason they are re-configured and take on new meanings. This is evident in differences through which Jewish communities reason about themselves and other Jewish communities, employing terms like progressive, traditional, and even primitive in discussing religious and cultural practice. Translation produces categorizations of Jewishness: as religion, culture, ethnicity and/or race. For example, "Israel" has been translated to a variety of categories, like nation, race, culture, and ethnicity that compete for authority both within and outside of Jewish communities.

Thus, hybrid systems of reasoning produced in negotiation between dominant and "other" systems of reasoning can be thought of as the site of translation. Yet paradoxically the process of translation always arrives at the point of its own impossibility: systems of reasoning cannot be reduced to one another and ultimately remain different. The "reformed, but different, other" challenges the universality of modern knowledge. It also remains necessary for continued comparative ordering and ranking of populations and ways of being that produce notions of modernity and progress. This process of mimicry that exposes "other" populations to, or imposes upon them, modern reasoning has effects of reforming and disciplining the "other," making him/her acceptable (Bhabha, 1994). Yet the "other" remains other and the object of comparative ranking and ordering, as modern reason shifts and notions of acceptable difference are re-configured. Processes of mimicry, translation, and reconfiguration of norms also "govern" individuals within the dominant population through enunciation and re-enforcement of the parameters of "normal" conduct.

In summary, cultural difference, as enunciated in translations of systems of reasoning, assigns relations of power. Constituted and stated as concepts like childhood, culture, race, and/or religion within standards of modern systems of reason, enunciations of difference mark parameters of inclusion/exclusion and assign power relations. Thus, enunciations of normal childhood constituted in discourses of child care quality are complicit in producing ranking and ordering of populations and individuals through notions of diversity which mark acceptable difference. The reasoning constituting child care quality and acceptable cultural difference also reconfigures "other" reasoning, such as that of "normal" Jewish childhood, as they are translated into "others" understandings of themselves. However, enunciations of difference continue to challenge universal norms, so while processes of translation (and mimicry) always promise possibilities for inclusion, enunciations of difference and sites of untranslatability delineate exclusions. Enunciations of Jewish difference continue to challenge modern universal norms—how can the Jewish child be both the same and different?

Modern Systems of Reason: Historically Constituting

Child Care Quality; Re-Configuring Jewishness

Having discussed the production of knowledges and differentiations that occur in the encounters between modern systems of reasoning I now turn to historical conditions through which child care quality has been constituted and to re-configurations of Jewishness that have occurred through related reasoning. Currently, discourses of quality have emerged as governing concepts across fields such as child care, health care, and business. This emergence across fields reflects shifts in reasoning about populations, desires, technologies of information and statistical analysis, and ideals of the future. While in-depth historical study of child care quality is beyond the scope of this chapter, I examine production of knowledges of childhood, social economy, and social movements to illustrate the normative, comparative, and constitutive effects of the reasoning.

Constituting Child Care Quality

Childhood

Assumptions about childhood as it emerged in enlightenment reasoning have been consistently re-inscribed within modern knowledges about child care. Rousseau's (1762/1974) metaphorical linkage of childhood with savagery reflects, and produces, the inscription of developmental progress— savagery to civilization—within notions of individual development. Thus populational comparisons privileging Western systems of reasoning inscribe the child as primitive and incomplete while inscribing "other" populations as childlike and irrational. Modern reasoning about the role of education shifted in accordance with new conceptions of knowledge and truth. Educational goals of ensuring individual redemption in the world to come through knowledge of scripture shifted to educational goals of redemption of the future in this world through development of capacities of reason.

In keeping with the notion of mimicry, different educational processes were considered appropriate for less rational and civilized populations. Few, other than the European, Christian (in the United States Protestant), male child were considered capable of fully developing the capacity for reason.[6] Therefore, while education was considered necessary to reform "other" populations, educational possibilities were often limited. In the United States where public education served to Americanize immigrants, participation of individuals from "other" populations was constrained through "expert knowledge" advocating vocational education for poor, immigrant, Indian, and Black children. Additionally, access to higher education for women and "others" was constrained due to their perceived lack of physical and mental capacity to develop reason and rationality.[7] Discourses of child care quality, drawing upon expert knowledge of child development are inscribed with assumptions of the child as different or lacking. They re-produce culturally inscribed norms of reason, civility and conduct against which individuals and populations are compared.

Social economy—Non-normal Populations

Assumptions about educational needs of children have also been affected by reasoning about non-normal populations. Discourses surrounding non-normal populations re-inscribed a redemptive concern for the future of society if children lacked a proper moral and civil education.[8] For example, notions of morality and civility were re-inscribed within reasoning surrounding poverty that stabilized and protected societies in the process of changing economic systems. Protestant redemptive beliefs have been re-inscribed within reasoning about poverty through binaries such as moral/immoral poor (later deserving/undeserving poor) allowing for categorization of individuals and populations along continua ranging from one end to another. This stabilized societies with changing economic systems by setting standards for governing conduct of the poor and obscuring the wealth/poverty binary (see Procacci 1987/1991).

Emerging knowledges about good parenting and educational needs of children from "other" populations have been re-inscribed with these notions of Protestant morality and individual deficiency. For example, the American Infant School movement of the early nineteenth century reasserted the inability of poor parents to provide adequate moral upbringing. However, as affluent children were enrolled in infant schools for supplemental education the benefits were challenged through medical claims of physical and emotional damage to young minds and bodies. This reasoning re-established normalizations of maternal care, special emotional needs of young children, and the differences between populations; subjecting both poor and affluent mothers to governing discourses about normal and non-normal childhood and motherhood. Poor children, with mothers assumed unable to provide adequate moral training, were constituted as needing basic moral education and vocational training. Affluent children, whose mothers were assumed capable of providing adequate care and education, were to be protected from mothers either overzealously pursuing early academics or too lazy to provide adequate care at home.

Differentiations between populations were re-inscribed in knowledges of child study and child development as they emerged at the turn of the twentieth century. Differential reasoning about educational needs of children produced, and was re-produced in, various institutions and disciplinary fields concerned with the care and education of children (see Bloch, 1992). Preschool or nursery education and supplemental parenting education based on knowledge of normal development were available to well-educated, financially secure families. However, charitable interventions and social policy were produced within, and re-constituted reasoning related to specific moral and cultural failings of less affluent (often ethnically and/or racially different) populations. The provision of full-time educational child care for all children would have challenged norms constituting normal conduct of family members and educators (see Michel, 1999).

Movements for Social Change Re-inscribing Difference
While Becoming Inclusive

Movements for political and social inclusion also re-inscribe and re-configure differentiations between populations. For example, the case for women's social and political inclusion at the turn of the twentieth century compared womens' civility and rationality to that of freed slaves and immigrant men re-inscribing racial and ethnic hierarchies. Additionally, the discourses re-inscribed gender by ascribing certain societal problems, like education and public health, as falling within parameters of women's "natural" interests and abilities (see Mink, 1990). Similarly, at the end of the twentieth century attitudes about non-maternal child care have been re-configured through reasoning asserting women's abilities to raise families and maintain careers. While attitudes towards non-maternal child care changed, guidelines for "quality" of care and "best practice" re-inscribed childhood as a delicate period with special needs. "Quality" of care emerged as a set of criteria or markers for differentiating child care environments, and in effect produced normalizations against which individuals and populations judge themselves and others (Kennedy, in press).

Re-configuring Jewishness

As modern (Western, Christian) systems of reasoning have changed, Jewish systems of reasoning have changed through encounters with it. In Europe, both Christian and Jewish communities had systems of reasoning in which god, truth, and representation were unified and reducible to each other. However, as small communities situated within larger dominant populations, Jewish refusal to acknowledge Jesus as the Messiah and the central authority of the Church in Rome led to violence toward, segregation of, and expulsion of Jewish communities. As previously noted, shifts in modern systems of reasoning affected both configurations of Jewishness within dominant reasoning and Jewish reasoning through the translations of Jewish systems of reasoning into "modern" systems of reasoning. Universal notions of individual agency emerging with shifts occurring in reasoning related to reformation, enlightenment, and secularization produced shifts in reasoning about Jewish participation in mainstream, dominant society. While Judaism remained a suspect religion, Jews became individuals to be reformed through education. The reform process differed within various European and non-European communities. In Western or Central Europe, the process was voluntary and Jewish individuals and communities participated in modern education systems producing good citizens and gaining broader access to employment and professions (although opportunities for higher education and occupational advancement remained limited). In Eastern Europe governments/populations often designated as "other" by their Western neighbors and struggling with their own encounters with modern reasoning, imposed modern secular education on Jewish populations while retaining strict limits on Jewish participation in society.[9] The secularly educated Jewish male was suspect in both dominant and Jewish communities, not being fully accepted as either one or the other.

While enlightenment reasoning held out the possibility of individual Jewish reform, Jewish populational differences were re-inscribed within new knowledges. Reconfigurations of Jewishness as racial, national, and/or ethnic called into question the possibility of real Jewish reform both reflecting and re-producing inscriptions of difference and foreignness. For example, overemotionality ascribed to Jewish men made them feminine in their inability to achieve rational civil-oriented thought (see, Gilman, 2003).

Additionally, rules of Christian theological reasoning were re-inscribed into populational differentiations. For instance, Church laws forbidding usury as a Christian profession created the "Jewish moneylender" by making it one of the few occupations in which Jews were allowed to participate. As rules of reason and systems of social economy changed, stereotypes and suspicions associated with the moneylender were re-inscribed in stereotypes of the Jewish banker. Thus, social economy was governed not only through differentiations between moral/immoral poor but through such differentiations as legitimate/illegitimate wealth.

Encounters between Western reasoning and Jewish reasoning resulted in a wide and varied range of hybrid forms of Jewishness. Multiple Jewishnesses emerged as forms of both resistance to, and acceptance of, modern reasoning, but always in translations of modern reasoning. This variety reflects differing and complicated encounters with dominant systems of reasoning.[10]

Jewish experience in the United States has differed from European and Eastern Jewish experiences. Decisions about Jewish settlement in North American colonies were often influenced by economic considerations which reveal and reinscribed the linkage of economic and religious differentiations in reasoning about Jews.[11] Additionally, unlike European nations where government and religion were often connected within a Church of State, multiple Protestantisms competed for authority in American colonies and no single one was recognized as an official religion of the State after the revolution. Thus, while religious difference remained an important factor in configuring populational difference in the United States, reasoning surrounding religious and populational differentiations took different forms and had different effects than in European contexts.[12] While remaining suspect, Jewish reasoning, mimicking practices of Protestant and secular mainstream reasoning without intervention or imposition by the government, produced multiple re-configurations of Jewishness. In the case of Jewish women, reasoning about their religious roles was reconstituted in various ways across North American communities, and while they differed from each other, there were greater differences with reconfigurations of discourses of Jewish womanhood in Europe. Overall, Jewish women in early national America were more likely to attend religious services, sit with family rather than in segregated pews, and participate in philanthropic pursuits than were European Jewish women.

Processes of hybridity reflected maintenance of Jewish belief and practice, while engaging shifting modern reasoning related to education and social welfare. Philanthropic pursuits incorporating expert knowledge of normal and non-normal populations affected Jewish notions of peoplehood and community responsibility. Hybridization is evident in the reasoning's mimicry of mainstream discourses of social welfare, but also in its resistance to the missionary proselytizing of Protestant charitable organizations (Hyman, 1995). Thus, in the 1800s, Jewish women philanthropists formed the Jewish Sunday School movement following dominant reasoning concerned with high rates of immigration and poverty but also emphasized Jewish education in resistance to conversion efforts of Protestant Sunday School activists. Not only did this movement both mimic and resist dominant social reasoning, but it also had effects upon Jewish reasoning about educational practices and gender roles, creating a new space for Jewish women to participate in community life.

In summary, current discourses of child care quality are inscribed with, and re-inscribe, cultural assumptions of civility, progress, and universal development already embedded within modern reasoning. These assumptions produce and inscribe "normalcy" and "populational difference" in thinking about children, their care, and their families. Within current reasoning, Jewishness may, or may not, fall into categories of acceptable difference. However, it is always somehow suspect as any enunciation of difference challenges the assumed universality of modern "truths." The encounter of Jewishness, or any "other" system of reason, with modern systems of reasoning produces shifts in both systems of thought. Dominant reasoning responds to enunciations of difference in the production of categories and classifications for translating this difference, while "other" systems of reason are translated and shift in the encounter. Thus, notions of Jewish childhood and family life have been, and continue to be, re-configured in relationship to modern systems of reasoning. At the same time, notions of Jewish childhood and family life as "the same but different" challenge established norms and has effects upon new reasoning and knowledges.

Child Care Quality: Governing Effects

Child care quality is situated within a grid of interacting, complementary, competing, fluid and contingent expert knowledge, practice, and commonsense. Each enunciation of quality (whether it is a formal scale of measurement, a critique of child care quality from within or outside the field, or the daily practice of a particular child care/education setting) produces differentiations between individuals and populations; ordering them in relationship to particular conceptions of normalcy. Hence, enunciations of child care quality mark the parameters of acceptable cultural diversity, re-inscribing as normal those cultural values embedded within modern reasoning.

In this section, I explore how "normal" is constituted through concepts of child care quality. I also examine how discourses of child care quality interact with enunciations of "other" cultural reasoning. I analyze configurations of "normal" childhood through two enunciations of child care quality that are as follows: (1) Quality care and normal childhood as constituted in a commonly used measure of quality, the Early Childhood Environment Rating Scale-Revised (ECERS-R) (Harms *et al.*, 1998), and (2) normal Jewish childhood as constituted through discursive practice in one Jewish preschool.[13]

As noted earlier, within modern reasoning knowledge is differentiated into objective, universal truths and particularistic personal beliefs. Quality as delineated in child development knowledge is assumed to represent universal human needs and universally valued outcomes. This is exemplified in the statement "Whatever the setting, it is believed that children require the same kinds of basic inputs for developmental success, although there is room for some flexibility in the details. Interestingly, the quality elements . . . appear to cross international borders" (Cryer, 1999 p. 49). "Crossing borders" suggests nationalities possess categorical differences. But ignoring the differences infers that national differences do not impact individual development. The effect is to privilege universal knowledges and relegate diversity to categories of personal belief categories like culture, ethnicity, and religion, which are understood as particularities "added-on" to the universal process of human development.

I explore assumptions about concepts of diversity and inclusion embedded, and produced, within the ECERS-R. Then I examine how these assumptions are translated into practice in the Jewish preschool. Diversity and inclusion are crucial aspects of the ECERS-R; a revision designed to address critique of omission of those issues in a previous scale. Interestingly, while aiming to address inclusion and diversity the ECERS series reifies differentiations and normalizations of developmental age and child care contexts through the existence of multiple alternative rating scales for different ages and settings.[14] Of course, developmental age is not a natural distinction but reflects and re-produces the inscription of the savage/civilized metaphor within modern reasoning. For example, Zborowski and Herzog describe Eastern European Jewish conceptions of the child at the turn of the twentieth century:

> Growing up is not a transition of the total child through graded phases. There is no nursery life in the shtetl, no kindergarten phase, no

recognized period of adolescence. Children mature in segments or
streaks, so that one segment may become comparatively adult while
another remains immature. (1962, p. 20)

This notion of childhood differs from age-defined stages of development
currently associated with early childhood education. It is difficult to tack
this very different conception of childhood onto a notion of developmental childhood. Yet, it is not necessarily clear that the Eastern European
Jewish notion of childhood is less progressive or evolved than current
developmental notions of childhood.

For many Jewish Americans this Eastern European notion of childhood
has interacted with, been challenged by, and has also challenged and
affected, shifting notions of normal childhood as produced through modern, objective knowledge and reason. Currently, some North American
Jewish early childhood educators are working to produce sets of age appropriate developmental guidelines for Jewish preschool education (Krug *et al.*,
2004). This process of mimicry produces categories, as well as hierarchies,
of "normal" Jewish childhood.

Diversity and Inclusion

The ECERS-R consists of 43 items grouped into seven subscales with
markers related to diversity woven throughout the measure. A very particular notion of diversity and inclusion emerges in the rating scale. Of 43
items, 22 include markers related to inclusion and diversity. In fact, there
are two items: one about culture (item 28) and one about (dis)ability (item
37) specifically focused on diversity and inclusion. Separate items related
to cultural/ethnic diversity, and (dis)ability constitute a double inclu-
sion/exclusion. They assert diversity and disability can and should be
addressed or included, but at the same time the reasoning sets them apart
from the "normal," and each other by constituting them as categorically
different types of diversity. Another example of a double inclusion/exclu-
sion is the item Meals/Snack (p.18). In order to receive a mid-level rating
(3.6) in this item children with (dis)abilities must be seated at the table
with peers. Thus, the inclusion of (dis)abled children in group activities,
like meals, is stressed. But on the other hand, setting (dis)abled off as a separate category of child re-inscribes the difference as non-normal. Why not
simply state that all children must be included at meals/snacks?

Throughout the ECERS-R diversity and (dis)ability are treated as
categorically different types of "inclusion" issues. Inclusion related to ability
(about 20 markers) receives more attention in the measure than inclusion
related to ethnic/cultural diversity (about nine markers). To understand
differences in the number of markers related to (dis)ability and culture
I explore three areas: expert knowledge, representation and inclusion, and
re-inscriptions of "normal" childhood.

Expert Knowledge

Many markers related to (dis)ability address staff solicitation of, and follow through on, expert knowledge related to disabilities. Multiple items note that staff members work with both experts and families. High ratings depend upon joint attendance of IEP (individualized educational program) meetings (p. 45), follow through on recommendations by other professionals (p. 45), and referring parents to other professionals when necessary (p. 46). The availability of expert knowledge related to (dis)ability makes it easier to normalize appropriate staff and parent actions and interactions around the nonnormal child. However, there is no suggestion in the ECERS-R to consult professionals about cultural difference, perhaps because it is difficult to identify appropriate experts or to translate differences into easily understood actions and reasoning.

Normalizations produced in child care quality discourses were dominant at the Jewish preschool. Teachers even felt the need to request professional advice for "including" Jewishness in the curriculum. However, due to the multiplicity of hybrid forms of Jewishness, obtaining cultural expertise was not a simple issue. In order to decide which experts to invite, the preschool community would have to define their Jewishness. Yet, an explicit enunciation of Jewishness would set boundaries upon "normal" Jewishness for the community and make participation less inviting for some families. Even within Jewish preschool, cultural or religious identity cannot be reduced to a simple list of characteristics added onto a modern, age appropriate curriculum.

Representation and Inclusion

Guidelines related to (dis)ability in the ECERS-R make interactions (often as interventions or fulfilling expert suggestions) with parents and children crucial to the maintenance of high quality. In contrast, there are few markers related to actual interactions with children of varying cultures. Most markers of cultural diversity have to do with representations of culture in the curriculum. High quality is attained through non-stereotypical representation in music, books, play materials, foods, and holidays. Of course, there are some markers defining appropriate interaction with children and their families—such as greeting children in their primary language. However, as this is a measure producing normalizations of childhood, difference is actually difficult. Thus the quote below, an addendum to an item on staff-parent interaction, rather than actually addressing difference, effectively reiterates 'normal' interactions.

> while the indicators of quality in this item generally hold true across a diversity of cultures and individuals, the ways in which they are expressed may differ. For example, direct eye contact in some cultures is a sign of respect; in others, a sign of disrespect. However, the requirements of the indicators must be met, although there may be some variation in the way this is done. (Harms, *et al.*, 1998, p. 40)

The issue of representation in achieving a score of high quality is also important when examining (dis)ability. Despite greater frequency of markers related to disability, it becomes a non-issue. There is only one marker that must be scored if there are no (dis)abled children enrolled. Hence, both diversity and inclusion are primarily dealt with as representation. Inclusion of culture and ethnicity in the ECERS-R are dealt with through representation rather than in interaction, and (dis)ability while discussed in terms of interaction is only an issue if there are disabled students enrolled. Hence, for both categories, representation becomes the primary means of addressing cultural or ability inclusion.

The Jewish preschool presented interactions with Jewishness through holiday ritual, Jewish songs, and some Hebrew language. Yet, the addition of "Jewish" interactions to a normal age-appropriate curriculum made "Jewishness" an "add-on" to normal childhood even in a Jewish preschool. In effect, Jewish childhood was constituted through modern categories and differentiations such that the preschool constituted a modern "progressive," rather than "traditional" Jewishness. This normalization of the modern, Western Jewish child establishes parameters that set multiple "other" Jewish childhoods outside the norm.

Some parents and experts, including myself (Bloch & Kennedy, 2000), suggested the preschool needed more representation of diverse cultures and emphasis on anti-bias curriculum. This had effects of further aligning the Jewish child with the normal child as it emerges within the quality literature. Interestingly, there was no suggestion to represent multiple Jewish cultures. Yet, multiple Jewish cultures could be represented through a variety of "modern" categories including race, ethnicity, or nationality rather than simply along a traditional/progressive (ranging from orthodox to reform) continuum of religious observance and practice. There were many unexplored categories through which to "add" Jewish difference to a curriculum that normalized a particular concept of modern Western Jewishness. Another approach would be to present multiple interpretations of ritual, practice, and language. Jewishness could be approached as multiple hybridities; interacting and translating modern reasoning about childhood in multiple ways and complicating simple mimicking of reasoning about diversity. Exploring multiple hybridities, making Jewish "the same, but different" within Jewish reasoning, has potential to trouble dominant discourses, normalizations, and categorizations.

Re-inscriptions of Normal Childhood Through "Inclusion"

Ranking and ordering are re-inscribed within child care quality discourses in two ways. First, re-inscription of wealth/poverty difference is obscured within notions of normal child development. Many markers of quality dealing with diversity/inclusion are specifically linked to material provision. For example, higher quality scores require many books and materials representing different cultures/genders/ethnicities in nonstereotypical ways. Material provision is particularly salient in markers related to (dis)ability which require adaptive materials and accessible physical space. However, as these are scored only if a (dis)abled child is enrolled it seems less expensive to maintain high quality by limiting enrollment.

Actually, markers of high quality, whether dealing with diversity/inclusion or not, are material dependent. Of 43 items, 33 require some type of material provision (even if it is simply space). High quality child care is expensive and, with the exception of tax credits, largely unsubsidized. Clearly, high quality child care, especially care that is inclusive and anti-bias, is expensive. Parental choice of high quality care is thus related to levels of affluence. Assumptions about wealth (moral/immoral, legitimate/ illegitimate) are re-inscribed in parental choice of child care. In effect, the constitution of diversity and inclusion within quality care discourses link tolerance, inclusiveness, and anti-bias curricula to economic and educational elitism. Thus, ranking and ordering of families by choice of care indicates the constitution of dispositions and "normal" attitudes toward "others."

The Jewish preschool untypically made use of donated games and toys from people's homes in addition to materials from professional catalogs. Our "expert" suggestion (Bloch & Kennedy, 2000) was to invest in better equipment and more space. The preschool had invested significantly in staff training and generous tuition assistance rather than always purchasing the most expensive materials. Our advice to invest in materials had potential effects on tuition cost, tuition assistance, or staff training and compensation. The potential effects of these suggestions would be the production of parameters of normal Jewish childhood dependent upon family income; mimicking modern reason and inscribing judgments about morality, intelligence, and "good" decision-making. A more expensive early childhood program might constrain participation of less affluent families. Distinctions in Jewish education and participation based on affluence could result in hierarchical and exclusionary practices within "progressive" Jewish settings. Additionally, it could unintentionally re-inscribe suspect notions of Jewish wealth within dominant reasoning; underscoring an underlying suspicion of deviance related to Jewish wealth embedded in reasoning of social economy.

The structure of programs also produces ranking and ordering within notions of "normal" childhood. Professional literature and research assume both the naturalness of maternal care and the importance of attachment. Thus, an important variable in the research, number of hours in non-maternal care, re-inscribes assumptions about both gender and affluence. The full-time working mother remains outside the norm. This is evident in the program structure of the Jewish preschool, as in the structure of many high quality programs. Many high quality settings offer educational morning or afternoon options with a variety of "wrap-around" programs providing additional hours of care. "Wrap-arounds" do not always have the same quality as the educational portion of the day. This creates a differentiation between supplemental education and custodial daycare and re-inscribes ranking and ordering of types of families utilizing different types of care.

While keeping staffing and material costs down the effect of "wrap-around" options is ordering of families by choice of care options that are often related to number of parents in the home and their working hours. So, while notions of "traditional" motherhood may be challenged by placing nonstereotyped materials in the classroom, the constitution of "normal" family structure and gender roles are re-produced through care and educational options differentiating between part-time educational supplement and custodial care.

This variety of program options also produces differentiations within the Jewish community and re-inscribes assumptions about Jewish difference within modern reasoning. The Jewish preschool offered multiple enrollment options with highly trained caregivers most evident in the educational morning program and various "wrap-around" options available for children with part- or full-time working parents. In general, these options were chosen by families with one parent who worked part-time and could not pick up children at the end of the "educational" part of the day but did not need daily full day care. These same options were not considered practical by single parents or families with two parents working full-time outside the home. The effects of this program structure were to constitute normal Jewish childhood through differentiations in family structure ironically re-inscribing assumptions about gender roles within a "progressive" setting. The mimicking of "normal" family structure re-produced within this notion of normal Jewish childhood does not, however, escape suspicions of Jewish difference. The active and involved Jewish mother risks perception as the stereotypical overbearing Jewish mother.

Quality child care emerges as an ordering of populations and individuals producing new differentiations and re-inscribing populational difference in terms of family structure and income. In translating these discourses of quality, Jewish preschools seeking to maintain Jewish knowledge produce a mimicking notion of Jewish childhood that ranks and orders Jewishness within modern comparative categories. Additionally, they are in danger of reaffirming suspect Jewish difference within those very categories.

Concluding Thoughts

This chapter is laid out in three interconnected parts. In the first section, I discussed modern knowledge of the child both as an effect of power and as having effects of power. As such, the knowledge produced both inscribes and is re-inscribed with differentiations between groups and people. In the second section of the chapter, I discussed historical conditions of possibility for current discourses of quality care and education of young children. As this reasoning emerged reasoning about Jews as "other" was reconfigured and reinscribed within new organizing principles.

In the third section of the chapter, I addressed the production of notions of diversity and difference within quality care discourses. Because modern

reasoning is both comparative and normalizing, issues of diversity and difference are seldom simple. As noted in the first section, the tension
between universality and the ranking and order of difference is a primary
conflict of modern reasoning. While difference is important and necessary
for comparative reasoning it is logically impossible to be both "different"
and completely "normal." To be "different," while making possible populational and individual comparison, also challenges understandings of universally applicable knowledge. This produces hybrid reasoning through
shifts, translations, and re-configurations of both dominant reasoning and
"other" reasoning. Yet, difference remains. This was clear in my exploration of diversity and inclusion as enunciated in both the ECERS-R and
in one Jewish preschool. Reasoning about diversity and quality care as
enunciated in the ECERS-R marked parameters of "normal" childhood
and differentiated between populations and individuals. The translation and
mimicking of these discourses in a Jewish preschool inscribed Jewishness
with these differentiations and simultaneously had potential effects of continuing to re-inscribe suspect notions of Jewish difference.

Currently, early childhood policy at the federal level calls for research on
quality practices in prereading and literacy skills. As this study of the effects
of quality discourses on configurations of Jewish childhood has shown, new
"quality" discourses will also reflect, re-produce, and re-inscribe notions of
"normal" childhood. These notions of "normal" childhood will mark differentiations between, and various inclusions/exclusions of, populations
and individuals. Additionally, the re-configured knowledge will affect and
shape how "others" seeking "inclusion" understand themselves. Not only
will notions of "others" be re-produced and re-configured, the "others"
will translate their understandings of themselves into new categories,
re-configuring and producing new differentiations within those "other"
populations.

Notes

1. *Good Start, Grow Smart,* http://www.whitehouse.gov/infocus/earlychildhood/earlychildhood.html
   proposes funding for programs like Head Start be tied to standards of accountability. Additionally,
   "quality" is highlighted through increased Federal funding for research and dissemination of
   findings.
2. See Memmi (1968) for an early postcolonial examination of processes of domination. See Boyarin
   (1992, chapter 5) for exclusionary aspects of spatial/racial reasoning of oppression.
3. See Mallory and New (1994), Cannella (1997) for critiques within the field; for a critique outside
   the field see Blau (1991).
4. Note that secular modern reasoning re-inscribes Christian "othering" of Jewish difference, albeit
   within reconfigured and new categories. See Anidjar (2003) for examination of reconstitutions of
   Jewish and Arab difference.
5. See Sutcliffe (2003) for description of encounters of Jewish reasoning with dominant Christian reasoning, shifts in Western reasoning associated with the Enlightenment, and effects on both
   "Western" (Christian) reasoning and Jewish reasoning.
6. Configurations of parameters of "secular" and "rational," and even "male" and "Christian" are fluid
   and contingent. These terms are not constituted today as they were in the Enlightenment.

7. "Other" populations have been configured in varying ways and have had differing experiences of exclusion. As there are multiple and varied categories through which norms of comparison are established, populations are configured differently within a variety of categories in modern reasoning and experience their "difference" in multiple ways. These differentiations inscribe both differences within groups and between groups.
8. See Donzelot (1978/1997) for a history of discourses of childhood surrounding normal and non-normal families.
9. For brief descriptions of Jewish encounters with "emancipation" and modernity see Birnbaum and Katznelson (1995). Jewish experiences with secular education and modern systems of reasoning were related to particular historical/cultural contexts in which they occurred. As non-Western populations were differentiated from Western populations, Jewish encounters with modern reasoning reflected complex relationships between "others." For reasoning about "Oriental Jews" see Rodrigue (2003) and Valensi (2002); for shifting within group reasoning about Jewish difference in Europe see Aschheim (1982).
10. See Aschheim (1982) for production of differentiations between Jewish groups emerging from Central European reasoning about charity and Jewish difference.
11. This reconfiguration of Jewish economic difference eventually had effects in Europe as Jews were granted permission to "settle" in England after they has "settled" in the colonies. See Sarna, 2004 for Jewish history in the U.S. context.
12. While primary populational differences (and racial categories) in Europe were constituted around religion/nationality, many scholars (e.g. Sarna, 2004), have noted the primary "difference" of concern in the colonies and United States was, and is, color.
13. Analysis of the preschool is based on a needs assessment performed for a local preschool by myself and M. Bloch a number of years ago.
14. See Infant/toddler environment rating scale (Harms, T. *et al.*, 1990); Family day care environment rating scale (Harms, T. & Clifford, R. M., 1989) and School-age care environment rating scale (Harms, T. *et al.*, 1996).

References

Anidjar, G. (2003). *The Jew, the Arab: a history of the enemy*. Stanford, CA: Stanford University Press.
Aschheim, S. E. (1982). *Brother and strangers: the East European Jew in German and German Jewish consciousness*. Madison: University of Wisconsin Press.
Bhabha, H. K. (1994). *The location of culture*. London: Routledge.
Birnbaum, P., & Katznelson, I. (1995). *Paths of emancipation: Jews, states, and citizenship*. Princeton, NJ: Princeton University Press.
Blau, D. (1991). *The economics of child care*. New York: Russell Sage Foundation.
Bloch, M. N. (1992). Critical perspectives on the historical relationships between child development and early childhood education research. In S. A. Kessler & B. B. Swadener (Eds.), *Reconceptualizing the early childhood curriculum: beginning the dialogue* (pp. 3–19). New York: Teachers College Press.
Bloch, M., & Kennedy, D. (2000). *The Madison Jewish community: A needs assessment related to early education and child care with a special focus on the Hilde Mosse Gan HaYeled Preschool*. Madison: University of Wisconsin Madison.
Boyarin, J. (1992). *Storm from paradise: The politics of Jewish memory*. Minneapolis: University of Minnesota Press.
Cannella, G. S. (1997). *Deconstructing early childhood education: Social justice and revolution*. New York: Peter Lang.
Carrette, J. R. (1999). Prologue to a confession of the flesh. In J. Carrette (Ed.), *Religion and culture: Michel Foucault* (pp. 1–47). New York: Routledge.
Chakrabarty, D. (2000). *Provincializing Europe: Postcolonial thought and historical difference*. Princeton, NJ: Princeton University Press.
Chakrabarty, D. (2002). *Habitations of modernity: Essays in the wake of subaltern studies*. Chicago, IL: University of Chicago Press.
Cryer, D. (1999). Defining and assessing early childhood program quality. *Annals of the American Academy of Political and Social Science, 563*, 39–55.

Donzelot, J. (1997). *The policing of families* (Robert Hurley, Trans.). Baltimore, MD: Johns Hopkins University Press. (Original work published 1978)

Foucault, M. (1972). *The archeology of knowledge and the discourse on language* (A.M. Sheridan Smith, Trans.). New York: Pantheon Books. (Original work published 1971)

Foucault, M. (1973). *The order of things: An archaeology of the human sciences.* New York: Vintage Books. (Original work published 1966)

Foucault, M. (1980). *Power/knowledge: Selected interviews and other writings, 1972–1977.* New York: Pantheon Books.

Foucault, M. (1991). Governmentality. (C. Gordon, Trans.). In G. Burchell, C. Gordon, & P. Miller (Eds.), *The Foucault effect: Studies in governmentality: with two lectures by and an interview with Michel Foucault* (pp. 87–104). Chicago, IL: University of Chicago Press. (Original work published 1979)

Foucault, M. (1995). *Discipline and punish: the birth of the clinic.* (A. Sheridan, Trans.). New York: Vintage Books. (Original work published 1975)

Foucault, M. (2003). *Society must be defended: Lectures at the College De France 1975–1976.* (D. Macey, Trans.). New York: Picador.

Gilman, S. L. (2003). *Jewish frontiers: Essays on bodies, histories, and identities.* New York: Palgrave Macmillan.

Harms, T., & Clifford, R. M. (1989). *Family day care environment rating scale.* New York: Teachers College Press.

Harms, T., Clifford, R. M., & Cryer, D. (1998). *Early childhood environment rating scale* (rev. ed.). New York: Teachers College Press.

Harms, T., Cryer, D., & Clifford, R. M. (1990). *Infant/toddler environment rating scale.* New York: Teachers College Press.Harms, T., Jacobs, E. V., & White, D. R. (1996). *School-age care environment rating scale.* New York: Teachers College Press.

Hyman, P. (1995). *Gender and assimilation in modern Jewish history: The roles and representation of women.* Seattle, WA: University of Washington Press.

Kennedy, D. (in progress). Quality child care: Assessing children, families, and caregivers. In S. Kessler & M. Bloch (Eds.). *Reconceptualizing the early education/child care curriculum: opening up to new possibilities.*

Krug, C., Feinberg, S., & Stevens, E.. (2004). *Defining excellence in early childhood education.* Medford, MA: Tufts University, Center for Applied Child Development Eliot-Pearson Department of Child Development.

Lipset, S. M. (2003). Some thoughts on the past, present and future of American Jewry. In S. Lyman (Ed.), *Essential readings in Jewish identities, lifestyles and beliefs: Analysis of the personal and social diversity of Jews by modern scholars.* New York: Richard Altschuler & Associates, Inc.

Mallory, B., & New, R. (1994). *Diversity and developmentally appropriate practice: Challenges for early childhood education.* New York: Teachers College Press.

Memmi, A. (1968). *Dominated man: Notes toward a portrait.* New York: Orion Press.

Michel, S. (1999). *Children's Interests/Mothers' Rights: The shaping of America's child care policy.* New Haven, CT: Yale University Press.

Mink, G. (1990). The lady and the tramp: Gender, race, and the origins of the American welfare state. In L. Gordon (Ed.), *Women, the state and welfare* (pp. 92–122). Madison: University of Wisconsin Press.

Popkewitz, T. S. (1998). The culture of redemption and the administration of freedom as research. *Review of Educational Research, 68* (1), 1–34.

Procacci, G. (1991). Social economy and the government of poverty (J. Stone, Trans.) In G. Burchell, C. Gordon, & P. Miller (Eds.), *The Foucault effect: Studies in governmentality: with two lectures by and an interview with Michel Foucault* (pp. 151–168). Chicago, IL: University of Chicago Press. (Original work published 1978)

Rodrigue, A. (2003). *Jews and Muslims: Images of Sephardi and Eastern Jewries in modern times.* Seattle: University of Washington Press.

Rose, N. (1990). *Governing the soul: The shaping of the private self.* London: New York: Routledge.

Rose, N. (1996). Governing "advanced" liberal societies. In T. Osborne, A. Barry, N. Rose (Eds.), *Foucault and political reason* (pp. 37–64). Chicago, IL: University of Chicago Press.

Rousseau, J. J. (1974). *Emile* (B. Foxley, Trans.). London: Dent. (Original work published 1762)

Sarna, J. D. (2004). *American Judaism: A history*. New Haven, CT: Yale University Press.

Sartre, J. P. (1948). *Anti-Semite and Jew*. (G. Becker, Trans.). New York: Schocken Books.

Sutcliffe, A. (2003). *Judaism and Enlightenment*. New York: Cambridge University Press.

Valensi, L. (2002). Multicultural visions: The cultural tapestry of the Jews of North Africa. In D. Biale (Ed.), *Cultures of the Jews: a new history* (pp. 887–932). New York: Schocken Books.

Zborowski, M., & Herzog, E. (1962). *Life is with people: the culture of the shtetl*. New York: Schocken Books.

CHAPTER 3

Problematizing Asian American Children as "Model" Students

SUSAN MATOBA ADLER

Introduction

Asian Americans[1] vary tremendously in their immigration histories, sense of ethnic identity and affiliation, degree of acculturation to mainstream U.S. societal norms, and amount of out-marrying by younger generations. I use the term *Asian American* as a collective term inclusive of a variety of ethnicities without an essentialized expectation of uniformity, but with some shared Asian cultural beliefs having roots in Confucianism. Despite the vast diversity among Asian Americans in the United States, as a panethnic[2] group, they have been historically constructed as homogenous and stereotyped as *model minorities* (Lee, 1996; Wu, 2002; Zia, 2000). This myth emerged in the 1960s from a sociopolitical context in which Asian American success was pitted against an African American deficit model. In this chapter, I discuss how this myth is still perpetuated in American society today, resulting in the need for Asian American parents to socialize their children to deal with racism and discrimination.

The purpose of this chapter is to use Homi Bhabha's (1994) notion of hybridity as a template for analyzing how Asian American children (Hmong in particular) negotiate Western (European American) and Eastern (Asian American) conceptions of "good school behavior" while others construct them (though they do not necessarily label them) as "model" citizens. I equate the assumption by school staff that Asian American students are passive and compliant in their behavior (traits associated with model minorities) to Bhabha's concept of *colonial identity*. Parents of minority Asian American students must begin to recognize public schools as a colonial space, which imposes the colonial authority of mainstream[3] European American norms and expectations. The model minority stereotype then becomes a method of colonization inhibiting Asian American students from appropriating their own identities.

I interpret students' struggle to construct themselves as Americans of Asian descent and to self-identify rather than being ascribed an identity by those in power as *hybridity*. Hybridity "reimplicates its identifications in strategies of subversion that turn the gaze of the discriminated back upon the eye of power" (Bhabha, 1994, p. 112). Hybridity is a relationship of what happens to the culture of the discriminated in negotiation of, for example, Asian American family expectations for behavior in school. For the dominant group, the identification of a model minority is not just a convenient differentiation between various "others" but it is also a disciplining technique ascribing stereotypic behaviors such as hard work, passivity, and obedience to Asian American students and policing them into a constructed homogeneous cultural category called the "model minority." For minority students, strategies of subversion might include challenging stereotypic behaviors by becoming assertive in covert rather than overt ways, eliciting collaborative support from other Asian American peers for asserting a culturally different position, and utilizing sociopolitical knowledge through the development of double consciousness to gain personal agency. Ladson-Billings (2000), in her chapter entitled *Racialized Discourses and Ethnic Epistemologies*, points out that Du Bois's notion of double consciousness is not one of a "pathetic state of marginalization and exclusion, but rather as a transcendent position allowing one to see and understand positions of inclusion and exclusion—margins and mainstreams" (p. 260). In order for minority students to "turn the gaze" back upon their colonizers, Asian American children need to learn from their parents and families how they have been marginalized.

In the chapter, first, I describe the diversity of Asian American families and some shared conceptions of the importance of children. Second, I provide perspectives and critique of the Model Minority Myth by Asian American scholars and teachers of Asian American students. Third, I raise issues about how Asian American parents socialize their children to develop identity, double consciousness, and resistance to discrimination and stereotyping. Finally, I analyze how Asian American hybridity and the model minority myth intersect in the identity formation of Asian American children with a caveat for parents, teachers, and administrators to seriously critique the issue of the colonization of Asian American students in U.S. schools.

Asian American Families

In my position as a university supervisor of preservice early childhood and elementary teachers and as a researcher of Asian American families in American schools, I have encountered many teachers of Asian heritage children who had no idea of their students' ethnicities.[4] When questioned about student ethnicities, a generic term "Asian" was commonly used by teachers to denote a racial category. But when asking parents or teachers of

Asian heritage, specific ethnicities (Japanese, Chinese, Korean, Filipino, Hmong) were the reported classification of identity. The following sections provide an overview of ethnic diversity among Asian American groups and a knowledge base for understanding shared themes on Asian American perspectives about children.

The diversity among Asian American families is so vast that even the category called "Asian American" has many definitions depending upon the use of the term. Asian Americans in the United States are a heterogeneous group of many ethnicities including Japanese, Chinese, Korean, Filipino, Asian (East) Indians, and Southeast Asians (Hmong, Cambodian, Vietnamese). They are not a single identity group, nor a monolithic culture; therefore, it is more accurate to speak of Asian American cultures (Zia, 2000). Early Asian groups were voluntary immigrants but after the Vietnam War, Southeast Asians were primarily refugees (Ng, 1998). The term Asian American identifies people with origins in at least 30 countries and also includes Amerasians (children of U.S. servicemen during the Vietnam War and their Indochinese mothers), adopted Asian children (a large number from Korea), and multiracial children of mixed Asian and European or African American marriages. Originally this mixed-heritage group included the children of war brides (wives of U.S. soldiers stationed abroad) primarily from Korea, Japan, and the Philippines. Currently it also includes children of interracial Asian American marriages and intra-ethnic Asian American marriages, more commonly found in places like California and Hawaii, where there are large Asian populations.

Family Structure and the Place of Children

The vertical family structure of patriarchal lineage and hierarchal relationships is common in traditional Asian American families, but there is diversity in practice across cultures (Yanagisako, 1985). Based on the teachings of Confucius, responsibility moves from father to son, elder brother to younger brother, and husband to wife. Traditionally, women are expected to be passive, and nurture the well being of the family. A mother forms a close bond with her children, favoring her eldest son over her husband (Hildebrand *et al.*, 2000). In the early 1900s, Japanese immigrant women were isolated, seeing other women only once a year. Consequently, they often became extremely close to their children (Chan, 1991). Thus, cultural tradition and living conditions fostered this close relationship. Over the generations, as in the case of the Japanese Americans, this pattern changed from the linear male oriented pattern of kinship to a stem pattern of shared responsibility and inheritance for both sons and daughters (Adler, 1998).

In contrast to the patriarchal and patrilineal structure of Japanese, Chinese, and Korean societies, the gender structure in the Philippines is more egalitarian and kinship is bilateral. In employment, women have and continue to have equal status with men, although employment occupations often differ (Espiritu, 1995). In the 1930s, Filipino families living in the United States maintained split-household families with fathers immigrating for work leaving their wives and children in the Philippines. Later, gender roles became reversed when Filipina women immigrated to the United States to become domestic workers and nurses in the health care system. They became breadwinners leaving their children and spouses in the homeland. Filipina women preferred having kin, rather than strangers, provide child care especially during infancy, even if that meant living away from their children. But this arrangement was considered a *broken home* since the ideal family was the nuclear family and there was an emotional cost of not being able to supervise one's own children.

Transnational (split-household) families grew out of economic necessity and transcended borders and spatial boundaries to take advantage of the lower cost of living for families in a developing country (Zhou & Gatewood 2000). Evelyn Nakano Glenn (1983) described Chinese split-household families as part production or income earning by men sojourning abroad and part reproduction or maintaining the family household, including childrearing and caring for the elderly by wives and relatives in China. Split-household families were common for Chinese between 1850 and the 1920s. These kinship patterns reinforced the cultural value of familism or mutual cooperation, collectivism, and mutual obligation among kin (Zhou & Gatewood 2000).

For Southeast Asian refugee families, the change in gender relations was a function of the changing gender roles upon relocation. Older men lost their traditional roles as elders who solved problems, adjudicated quarrels, and made important decisions, when they became powerless without fluency in English and understanding of Western culture (Chan, 1994). Non-English speaking Hmong parents needed their children to serve as cultural brokers, which undermined the father's traditional patriarchal authority, thus reducing his status. Hmong women, on the other hand, often being non-English speaking, discovered that they had more rights and protection from abuse as immigrants which made their adjustment somewhat easier than their husbands'. In addition, they sold their intricate needlework, providing family income (Chan,1994). Consistent with other Asian heritage groups, Hmong families value children and invest heavily in their futures. Today, contemporary Hmong parents, who have two (or more) incomes and little time for recreation, often use grandparents and other fictive kin as caregivers and linguistic guides for their young children (Adler, 2004).

There are a variety of reasons for the creation of extended family households, including collective care of children, support of parents and grandparents, the inculcation of language and culture, economic stability, cultural obligation, and family reunification patterns. Families living in the same household resulted from cultural norms, economic needs, and the process of migration. I offer three different examples of extended Asian family households. First, discrimination in housing, and economic necessity after World War II often brought a variety of Japanese family (and non-family) members together in one household. In addition, older Issei,[5] who could not speak English, relied upon their children, the Nisei,[6] to help them negotiate daily living in mainstream society (Adler 1998; Takaki, 1989). Households might include parents, children, unmarried siblings, and grandparents. Second, Asian Indian families believing in their traditional family structure reside as a joint family, which includes a married couple, their unmarried children, and their married sons with their spouses and children. Thus, three or more generations may live in the same household sharing space and responsibility for childrearing (Bacon, 1996). And third, Hmong families settled in the United States might incorporate relatives or clan members as they emigrated from refugee camps in Southeast Asia into their households, thus providing support for acculturation into American society (Adler, 2004).

Religion and a Collective Orientation

Many Asian heritage groups share commonalities having roots in Confucianism and other Asian cultural traditions such as placing high value on education and the family (elders and children). For example, Japanese children until age seven are considered to be a gift bestowed by God, therefore children are cherished and thought to be inherently good (Hendry, 1986; Yamamura, 1986). There is no equivalent for the Western concept of original sin or rebellious spirit. Sanctification of children has diminished, yet the belief in the essential goodness of children has not changed (Adler, 1998). It is important to recognize ethnic group diversity since many Asian Americans may maintain Asian cultural norms along with Western religious belief. They often live in a bicultural or multicultural world.

Asian immigrants arrive in the United States with many religions including Buddhism, Confucianism, Hinduism, Islam, and Christianity. The kinds of interpretive frameworks provided by religion, as a central source of cultural components, become particularly important when people are coping with changing environments (Zhou & Gatewood, 2000). Immigrants make sense out of their new environment by utilizing cultural components from traditional religion and subtly altering them to reflect the demands of the new environment. The diversity of Asian immigrant religions include Theravada Buddhism, Mahayana Buddhism, Shamanism (Hmong), Christianity (primarily Koreans and Filipinos), and forms of Hinduism, Sikhism, Jainism, Sunni and Shi'a Islam, and Syro-Malabar Catholicism (primarily Asian Indian and Pakistani) (Zhou & Gatewood, 2000).

It is through organized religion and family modeling that values and beliefs are inculcated to the younger generation. Although there are distinct differences among the Asian ethnic groups, there are some commonalities in worldview such as the following: group orientation (collectivity), family cohesion and responsibility, self-control and personal discipline, emphasis on educational achievement, respect for authority, reverence for the elderly (filial piety), the use of shame for behavioral control, and interdependence of families and individuals (Hildebrand *et al.*, 2000).

Group collectivity, accepted by most Asian cultures, is illustrated in the Asian Indian religious perspective. In the East Asian Indian worldview there are no individuals, rather, each person is born with a distinctly different nature or essence, based upon his or her parents and the specific circumstances of birth. This makes people fundamentally different, rather than the same (or equal), and this nature changes over time (Bacon, 1996). This holistic worldview ties Asian Indian identity to social relationships and the inherent inequality gives rise to social rankings based upon social relations. The caste system can be visualized as a system of concentric circles in which the social groups that encompass others are ranked higher than those they encompass, rather than a ladder system of inequality. As a result of this traditional worldview, Asian Indians regard social relationships as the building blocks of society. Children learn their place within the sociocultural and gendered structure of their families and communities. In contrast, a "Western" perspective puts individual choice as the foundation of group affiliation and children are taught to assert themselves among peers (Bacon, 1996).

The Model Minority Myth

The need to problematize Asian American children as "model" students arises from the dilemma students of color face being ascribed a racial/ ethnic identity rather than having agency to appropriate their own identity. Identity formation is based upon a variety of cultural and experiential factors such as heritage, language, family beliefs and traditions, and degree of acculturation into mainstream American society. Thornton (1992) writes:

> To have identity means to join with some people and depart from others; it is a dialectic between identification by others and self-identification, between objectively assigned and subjectively appropriated identities. Much of an individual's inner drama involves discovering the assigned identity and recognizing that certain groups are or are not significant (p. 173).

As with most ethnic immigrant groups in American history, Asian Americans, as a group, have been subjects of stereotypes, or group definition by others, depending upon the sociopolitical context of the time. Early stereotypes of immigrants described Japanese and Chinese as *Orientals* who could not be assimilated. Then, during wartime hysteria, Japanese and Japanese Americans were characterized as the *yellow peril* (although this label had been prominent since their arrival in the 1800s) and any Asian in the United States was still considered a *perpetual foreigner*. Postwar years and the impact of higher education on Asian Americans brought the stereotype of the over achieving, *model minority* (Chan, 1991). Geishas, gooks, and geeks have been the major staple of Asian stereotypes with men portrayed as untrustworthy, evil, or ineffectual, emasculated nerds, and women cast as subservient, passive females, or seductive, malicious dragon ladies (Zia, 2000). Although stereotyping clearly remains, the desire to be "politically correct" and not to offend minorities, has tempered the overt expression of group labels.

Critique of the Model Minority Stereotype

So why is this "Model Minority" stereotype a myth? In his book *Yellow: Race in America beyond Black and White*, Frank Wu (2002) asks the questions: "Model for what?" And "Model for whom?" Wu suggests that this constructed phenomenon can be interpreted in two ways: (1) as implying that Asian Americans are remarkable, given their minority racial group standing, or (2) that they are "model" at least for people of color, satisfying a lesser standard (p. 59). When compared within a school setting, Asian American children can be compared with mainstream European American middle class, often suburban students or with minority racial/ethnic groups, often from urban settings, such as Hispanic and African American. In either setting, Asian American students are constructed as "high achievers" and "hard workers" (Adler, 1998, 2001, 2003).

Lee (1996), in her book, *Unraveling the "Model Minority" Stereotype: Listening to Asian American Youth* describes the stereotype's connection to claims of inequality:

> In all its permutations, the model minority stereotype has been used to support the status quo and the ideologies of meritocracy and individualism. Supporters of the model minority stereotype use Asian American success to delegitimize claims of inequality made by other racial minorities. According to the model minority discourse, Asian Americans prove that social mobility is possible for all those who are willing to work. Asian Americans are represented as examples of upward mobility through individual effort (p. 8).

Why is this stereotype so pervasive and why won't it disappear? Wu (2002) explains this question in the following way:

> The model minority myth persists, despite violating our societal norms against racial stereotyping and even though it is not accurate . . . The myth has not succumbed to individualism or facts because it serves a purpose in reinforcing racial hierarchies. Asian Americans are as much a "middleman minority" as we are a model minority. We are placed in the awkward position of buffer or intermediary, elevated as the preferred racial minority at the expense of denigrating African Americans (p. 58).

The middleman position is a good example of mimicry; holding out the possibility of becoming an acceptable part of society, while at the same time, reifying societal norms by reaffirming difference in a way that is not necessarily accepting of difference.

Teacher Perspectives on the Model Minority Stereotype

As a University supervisor of Japanese heritage, I recall the day that I visited my son's school (He was a 3rd grader at that time) with a group of education students who were observing classrooms. While meeting one of my son's teachers in the hall, I was told how respectful my son was and how the teacher enjoyed having the "Asian" children in her class because they were so "well behaved." When I asked what the ethnic backgrounds of her Asian heritage students were, she paused, then indicated that she "didn't have a clue." "Does it matter?" she asked. I quickly commented that there was much diversity in the Asian cultures and pointed out that my son was adopted and half Filipino, even though I was 100 percent Japanese American. That was in the early 1990s and in a 2004 research study I asked elementary teachers a similar question (see Adler, 2004).

In my qualitative study on Hmong parent involvement in schools (Adler, 2004) I asked staff what they thought about stereotypes of Asian Americans and to describe their Asian American students (the elementary school was over 50 percent Southeast Asian). There was a real divergence of views with descriptive words that were consistent with the model minority stereotype, such as "obedient," "well behaved," "hard working," and "respectful." There also appeared to be a desire to represent themselves and their students in a positive (and perhaps "politically correct") light to an Asian American researcher. When asked to comment on stereotypes, one English Language Learner (ELL) teacher described the model minority as applied to Southeast Asians in the following way:

> The "Model Minority" stereotype is not necessarily a valid assumption in [this area] when applied to the South East Asian population. There has been a great deal of media coverage of gang related violence and murder . . . regarding the Hmong population, so many people have a stereotype of the Hmong as violent gangsters. Also, when the Hmong first arrived, many of them received welfare. So they are also seen as illiterate welfare recipients. Parts of both are true. As in the fact that their culture does teach respect for elders. So, it has been my experience that some students are passive and quiet in class and try very hard. You can also identify some students who have gang affiliations, and their behaviors follow suit. And students are often poor- although I have seen a rapid rise in outward appearances of higher incomes: improved clothing, availability of money for book orders, etc. (Adler, 2004)

As you can see from the previous description, this stereotype is problematic and cannot be easily applied to all Asian American ethnic groups. Thus, cognizance of the diversity within any racial group and the diversity among various ethnic groups is vital for teachers working with students in today's multicultural, mixed-heritage, and linguistically diverse American society.

The Socialization of Asian American Children

Children of Asian heritage in the United States develop their sense of racial/ethnic identity from the cultural messages of their families and by messages from people, texts, media, and peers in their environment. Kitano and Daniels (1988) theorized that the dominant mainstream society, which provided employment and thus livelihood for immigrant families, is the primary socializing force while the ethnic community and families provided "supportive frameworks" (p. 73). Referring to Japanese Americans they wrote:

> Sansei[7] are more apt to reflect the ambience of their surrounding communities, rather than a strictly ethnic one. A Japanese American growing up in St. Louis will be more Missourian than Japanese, just as Sansei from Los Angeles, Honolulu, and New York will reflect the culture of these cities (p. 73).

Young children are socialized by their family's values and beliefs until adolescence when they begin to make conscious choices about group affiliations and identity (Phinney, 1989). For some, racial visibility is more salient than ethnic identity, especially when ethnicity is interpreted as *Asianness* or foreign culture; for others, a desire to maintain family ethnic culture is more meaningful. In any case, identity is both ascribed by peers and community and appropriated by personal choice (Thornton, 1992). Racial awareness may also lead to panethnicity, or feelings of affiliation with other Asian Americans, which is based upon commonalities in Asian culture and the need for political agency as a group (Espiritu, 1992).

Ramsey (1991) developed *stages of racial attitude development* for young children that can be a helpful guide for parents and teachers to respond to children in age-appropriate ways. Preschoolers (3–5 years) would have a rudimentary concept about race and may repeat but not understand comments made by others about race. Early elementary children (5–7 years) can clarify which physical characteristics belong to a particular race and learn that racial characteristics are permanent in people. Middle elementary children (7–10 years) develop concepts and feelings about their own and other races. They crystallize attitudes about race and reflect viewpoints of their families and communities (p. 55). Research also indicates that children of color are aware of race at earlier ages than their White counterparts (Holmes, 1995).

How Asian American parents socialize their children about race and ethnicity is often a function of their own upbringing. In the Midwest, that often meant little contact with ethnic communities and facing differing degrees of cultural alienation and discrimination. One Chinese American participant in my Detroit Study[8] described his feelings growing up:

> We were the only Chinese family in the area . . . My parents just ignored everything Chinese . . . I did not have that many Asian friends to hang out with. My parents would not send us to Chinese school but wanted us to go to Chinese church. I remember that people made fun of me because I looked different.

This Chinese American father married a woman from China and was actively raising his three-year-old son to be bicultural and bilingual. Discrimination taught him that race was a salient factor to consider as a parent and he chose to give his son a stronger sense of ethnicity.

Another example of a common parental response when dealing with issues of identity came from a Filipino mother who recalled being in a restaurant where there were no other people of color. Her children asked why some European American children were staring at them. Later, in the restroom, her daughter was asked if she could speak Chinese. "What did you tell her?" inquired the mother. "I said no, but that I was Filipino and went to Filipino school as well as American school," replied the daughter. Though the mother was uncomfortable with a potentially offensive situation, her seven-year-old daughter attached no sense of racial or ethnic prejudice to the incident.

These are just two examples that illustrate the complexity of socializing children to understand issues of race and ethnicity and to embrace positive racial/ethnic *identities*. Moreover, the development of a double consciousness and learning how to negotiate multiple cultural contexts can be overwhelming for some parents of racial minority children. One common approach is to teach children to accept a colorblind perspective and like the Chinese father's parents, ignore race. Asian American parents who had experienced name-calling and stereotyping throughout their lives, advised their children to ignore the comments, or to rise above them by being better (wiser, stronger, smarter) than their tormentors. Some Asian American parents did not discuss prejudice and discrimination directly with their children, though it was acknowledged as part of life. Children were expected to endure and persevere, which would make them mentally stronger (Adler, 1998).

I find this approach to be counterproductive because it does not allow Asian American children to develop strategies for challenging racism. Another perspective to challenging discrimination and racism is for racial minority children to develop a strong self-concept. One Japanese American mother explained that when she was in middle school, her father told her to "just remember that you are smarter than those who make fun of you" implying that her academic prowess would shield her, at least psychologically, from racism (Adler, 1998). She passed this message on to her own children. Although this approach may be a defense mechanism, it does not allow Asian American children to develop and appreciate a double consciousness, nor does it help them build strategies for resistance. Asian American parents need to socialize their children to challenge the ascribed stereotype of the "model minority" and to appropriate their own identities.

Asian American Hybridity and Resistance

When I think back to the previously discussed example of a teacher describing all of her Asian American students as respectful and well behaved, I ask "*Who* has the 'right' to label and classify children based upon *what* ethnic or cultural traits assumed to be 'true'?" How did the historical evolution of this model minority stereotype become the template by which educational professionals signify desired behavior by which Asian American students are measured and compared?

Perhaps Asian American parents and children construct their own identities in ways that traverse different boundaries and reflect their lived experiences from bicultural or multicultural perspectives. Perhaps the attributes that teachers use to describe their Asian American students are the same words but with different cultural meanings and contexts than Asian American parents use to inculcate their values. For example, does the phrase "respect elders" (or respect authority) mean "to hold an individual in high regard" or "to honor a higher position in a hierarchical system?" I suspect that Asian American students show respectful behavior toward their teachers because of the latter but do mainstream teachers assume the former? Bhabha (1994) writes:

> The representation of difference must not be hastily read as the reflection of *pregiven* ethnic or cultural traits set in the fixed tablet of tradition. The social articulation of difference, from the minority perspective, is a complex, on-going negotiation that seeks to authorize cultural hybridities that emerge in moments of historical transformation. The "right" to signify from the periphery of authorized power and privilege does not depend on the persistence of tradition; it is resourced by the power of tradition to be reinscribed through the conditions of contingency and contradictoriness that attend upon the lives of those who are "in the minority" (p. 2).

In this case, I interpret the "*pre-given* ethnic or cultural traits" as attributes attached to the model minority stereotype such as "hardworking," passive, and compliant. And I raise the question, when and how do minorities being ascribed this stereotype resist and challenge this "right to signify" by the majority? How do Asian American parents socialize their children to resist such labeling (and its implied expectations)? How do Asian American students resist ascribed identity by mainstream teachers (based upon their knowledge and personal biases) and seek to appropriate their own identities based upon multiple factors such as biological and fictive kin, nuclear and extended family, and ethnic community? The cultural hybridities that Bhabha describes are realities for ethnic minorities like Asian Americans and call upon us to *reinscribe* our own multicultural, and for mixed-heritage children, biracial or multiracial identities. In describing resistance to discrimination and domination, Bhabha (1994) writes:

> Hybridity is the revaluation of the assumption of colonial identity through the repetition of discriminatory identity effects. It displays the necessary deformation and displacement of all sites of discrimination and domination. It unsettles the mimetic or narcissistic demands of colonial power but reimplicates its identifications in strategies of sub-version that turn the gaze of the discriminated back upon the eye of power (p. 112).

The following vignette illustrates how complex resistance can be and how "strategies of subversion" might be played out in the classroom. It is called *My Encounter with Sammy, the Hmong First Grader.*

> I was observing a summer session for Hmong and other students needing extra instruction in an urban school district where I had been conducting research. I agreed to supervise the group of Hmong children while my translator took her daughter, who was assisting in her mother's class that morning, to work. I needed to watch the children for 30 minutes, making certain the ones leaving joined the bus line, and to acknowledge the arrival of parents for our group interview session. Armed with a popular children's video, I sat the children down in the library and stereotypically assumed they would all behave and pay attention. After all, Sammy was the son of one of the Hmong teachers, and I had observed him and other "obedient" Hmong children in the hallways and classrooms under the supervision of Hmong teachers.
>
> Thinking that I had much to my advantage, a Disney video, years of experience as an EC/elementary teacher, and after all, I *was* Asian American too, I proceeded with my short term assignment. To my surprise, Sammy immediately began to challenge me (a common behavior for some children his age) by lying down, touching his neighbors and actively trying to get my undivided attention. And of course, the other children began to pay attention to Sammy and I had to make some quick "teacher moves" to refocus the class. I switched gears, got the children to the bus, and began to greet the parents and was thankful when my translator (and accomplished teacher) arrived.

Was this an example of beginning "strategies of subversion" and resistance by age, gender, and Asian-American hybridity, or was it just a cute, lively, intelligent boy exerting his personality at the end of a busy morning? Was Sammy (or his parents) aware that others construct him (an Asian American) as well behaved, hard working, and compliant and hold him to high expectations? There are many reasons why Sammy did not respond to me as he would his Hmong parents and teachers, and why he chose to challenge my "authority" at that time. Resistance to inequities is often discouraged, especially in Asian heritage groups such as the Japanese Americans who faced overt discrimination and internment during World War II and Hmong refugees who were forced to leave Southeast Asia during the Vietnam War. Diversity among ethnic groups may be a function of differing historical contexts, specific ethnic group norms and traditions, and varying interactional styles that may or may not lend themselves to expressions of resistance. It is therefore critical that researchers consider the influence of racial identity, racial/ethnic socialization, and perspectives on power relations within mainstream schools in their interpretations of minority student adjustment and achievement.

Schools as Colonial Spaces

Bhabha (1994) also speaks of the transformation of knowledges of cultural authority to include *native* knowledges. He writes:

> Culture, as a colonial space of intervention and agonism, as the trace of displacement of symbol to sign, can be transformed by the unpredictable and partial desire for hybridity. Deprived of their full presence, the knowledges of cultural authority may be articulated with forms of "native" knowledges or faced with those discriminated subjects that they must rule but can no longer represent . . . Such a reading of the hybridity of colonial authority profoundly unsettles the demand that figures at the centre of the originary myth of colonial power (p. 115).

Mainstream American public schools can be interpreted as sites of colonial space with colonial authority in the hands of school teachers, administrators, policies, and practices. In the case of Asian American families, one common panethnic belief, the "Eastern" norm of placing family responsibility above individual desire, could be interpreted as a "native" knowledge which conflicts with a more "Western" mainstream focus on individuality. Asian American children have to learn to negotiate cultural differences and seldom discuss this difficulty of living biculturally multiculturally with their parents, though most parents subtly help children to make these negotiations if they have had to deal with them themselves (Adler, 2003). Depending upon the parents' awareness of cultural difference or "cultures

of differences" (Ellsworth, 1997, p. 20) and their own degree of acculturation, the ability for offering support and advice can be limited.

Teachers also are not necessarily cognizant of "native" knowledge of
their Asian American students and can miss cues of cultural misunderstand-
ings especially if they construct their Asian American students as "model
minority." Young children (preschool and elementary aged) adapt easily by
communicating nonverbally with peers and copying behavior they
observe. These approaches are commonly adopted by Asian American students, especially those who are second language learners and who have
been inculcated to seek harmony in interpersonal relations. Asian American
interactive styles such as placing priority on "quiet" learning rather than
discussion and active questioning within the family context can be interpreted by educators as compliance and "good" desirable classroom behavior
or as quiet passivity.

Conclusion

Any racial, ethnic, or cultural group defines its educational norms based
upon its cultural beliefs (social and religious) about family and children, the
value placed on learning as a means to good citizenship, and respect for the
institutions where education takes place. The nature of the mainstream
school culture and the beliefs of staff working within the system play
important roles in determining individual student success. In the case of
minority populations, a discontinuity or lack of congruence between family
and school worldview causes a power differential between two possibly
different value systems. Often this power differential is invisible because
school personnel may be unaware of cultural differences (as in the case of a
belief that the "model minority" share similar cultural beliefs and behavior
with the majority middle-class Euro-American culture in the United
States) and minority parents and families are not empowered to articulate
and contest the ensuing problems leading to student failure (or discontent).

Parents of minority Asian American students must begin to recognize
public schools as a colonial space, which imposes the colonial authority of
mainstream norms and expectations. Rather than socializing their children
to assimilate while risking the loss of their native language and culture, they
need to work with administration and staff to meaningfully resist that
which is imposed by the institution and subject to the beliefs and biases of
those who work within those schools. Parents and educators need to work
together to facilitate the process of helping students appropriate their own
identities rather than accepting stereotypes, such as the model minority,
ascribed to a racial/ethnic group. Asian American parents need to inculcate
a strong bicultural/multiracial sense of self-identities in their children, making certain that cultural difference and "cultures of difference" are articulated to school personnel and that both ethnic group as well as individual
agency is maintained. This is not an easy task and will require collaboration

in order for institutions in U.S. society to become more multicultural and inclusive of Asian Americans as "insiders" rather than "perpetual foreigners" or "model minority."

Finally, while the concept of "insiders" may imply assimilation to the "model" or to form of "normal," the notions of multiple identities and a history of colonial difference, racialized hybrid cultures in the United States cannot be erased (Young, 1990). Therefore, we may expect that modern schooling represents a discourse of inclusion while exclusions will continue. Multiple cultures of difference will continue to exist *within* individuals, across and within homogenized ethnic/racial groups, and within and across nations more generally.

Notes

1. Asian American is nonhyphenated to denote Asian as a descriptor of Americans having Asian heritages.
2. Panethnic or panethnicity is described by Yen Espiritu as a political term to reflect a shared history of racial discrimination and prejudice in the U.S.
3. Mainstream American schools maintain curriculum based upon European American child development and child psychology norms and pedagogy that addresses needs of children from European American cultural backgrounds.
4. Ethnicity refers to the Asian country of origin and the cultural practices and beliefs of a subgroup of Asian Americans.
5. Issei are immigrant Japanese or first generation in the U.S.
6. Nisei are second generation Japanese Americans born in the U.S.
7. Sansei are third generation Japanese Americans born in the U.S.
8. Participants in my Detroit study on racial or ethnic socialization included Chinese, Japanese, Korean, Filipino, and Hmong families.

References

Adler, S. M. (1998). *Mothering, education, and ethnicity: The transformation of Japanese American culture.* New York: Garland Publishing, Inc.

Adler, S. M. (2001). Racial and ethnic identity formation of Midwestern Asian American children. *Contemporary Issues in Early Childhood, 2* (3). Retrieved February 14, 2003, from www.triangle.co.uk/ciec

Adler, S. M. (2003). Asian American families, USA. In J. J. Ponzetti (Ed.), *International encyclopedia of marriage and family relationships* (2nd ed.). (pp. 82–91) New York: Macmillan.

Adler, S. M. (2004). Hmong school community relations. *The School Community Journal, 14* (2), 57–75.

Bacon, J. (1996). *Life lines: Community, family, and assimilation among Asian Indian immigrants.* New York: Oxford University Press.

Bhabha, H. K. (1994). *The location of culture.* New York: Routledge.

Chan, S. (1991). *Asian Americans: An interpretive history.* New York: Twayne.

Chan, S. (1994). *Hmong means free: Life in Laos and America.* Philadelphia, PA: Temple University Press.

Ellsworth, E. (1997). *Teaching positions: Difference, pedagogy, and the power of address.* New York: Teachers College Press.

Espiritu, Y. L. (1992). *Asian-American panethnicity: Bridging institutions and identities.* Philadelphia, PA: Temple University Press.

Espiritu, Y. L. (1995). *Filipino American lives.* Philadelphia, PA: Temple University Press.

Glenn, E. N. (1983). Split household, small producer and dual wage earners: An analysis of Chinese-American family strategies. *Journal of Marriage and Family, 45,* 35–46.

Hendry, J. (1986). *Becoming Japanese: The world of the pre-school child*. Honolulu: University of Hawaii Press.

Hildebrand, V., Phenice, L. A., Gray, M. M., & Hines, R. P. (2000). *Knowing and serving diverse families* (2nd ed.). Columbus, OH: Merrill.

Holmes, R. M. (1995). *How young children perceive race*. Thousand Oaks, CA: Sage

Kitano, H. L., & Daniels, B. (1988). *Asian Americans: Emerging Minorities*. Englewood Cliffs, NJ: Prentice-Hall.

Ladson-Billings, G. (2000). Racialized discourses and ethnic epistemologies. In N. K. Denzin & Y. S. Lincoln (Eds.), *Handbook of qualitative research* (2nd ed.). Thousand Oaks, CA: Sage, 257–278.

Lee, S. J. (1996). *Unraveling the "model minority" stereotype: Listening to Asian American youth*. New York: Teachers College Press.

Ng, F. (1998). *The history and immigration of Asian Americans*. New York: Garland.

Phinney, J. S. (1989). Stages of ethnic identity development in minority group adolescents. *Journal of Early Adolescence, 9*, 34–49.

Ramsey, P. G. (1991). *Making friends in school: Promoting peer relationships in early childhood*. New York: Teachers College Press.

Takaki, R. (1989). *Strangers from a different shore: A history of Asian Americans*. Boston, MA: Little, Brown and Company.

Thornton, M. (1992). Finding a way home: Race, nation and sex in Asian American identity. In L. C. Lee (Ed.), *Asian Americans: Collages of identities* (pp. 165–174). Ithaca, NY: Asian American Studies Program, Cornell University Press.

Wu, F. H. (2002). *Yellow: Race in America beyond black and white*. New York: Basic Books/ Perseus Book Group.

Yamamura, Y. (1986). The child in Japanese society. In H. Stevenson, H. Azuma, & K. Hakuta (Eds.), *Child development and education in Japan* (pp. 28–38). New York: W.H. Freeman & Company.

Yanagisako, S. J. (1985). *Transforming the past: Tradition and kinship among Japanese Americans*. Stanford, CA: Stanford University Press.

Young, R. (1990). *White mythologies: Writing history and the West*. New York: Routledge.

Zhou, M., & Gatewood, J. V. (2000). *Contemporary Asian America: A multidisciplinary reader*. New York: New York University Press.

Zia, H. (2000). *Asian American dreams: The emergence of an American people*. New York: Farrar, Straus, and Giroux.

PART 3

Governing the Modern and Normal Child through Pedagogical Discourses

CHAPTER 4

Language Learning, Language Teaching, and the Construction of the Young Child

THEODORA LIGHTFOOT

They said I should learn to speak
a little bit of english
don't be scared of the suit and the tie
learn to walk in the dreams of the foreigner.
> (From Third World Child by Johnny Clegg and
> Savuka, cited in Pennycook, 1994, p.1)

Can we understand first and second language learning among young children in a way that is both scientific and culture free? Both scholarship and teacher education in the fields of foreign and second language teaching are strongly grounded in the scientific theories of linguistics and cognitive psychology. Language learning and language teaching are frequently presented as "knowable" processes about which the truth can be discerned from two sources. One of these sources is direct observation of the process of teaching and learning foreign and second languages. The second source of information, however, often considered more "pure" than classroom observations of second language learning, is theoretical and empirical understandings of how young children learn their first languages as infants and as toddlers.

First language acquisition is often considered a gold standard for understanding other types of language learning, for two reasons. Because many cognitive scientists regard culture as a complicating variable in their efforts to understand how people learn and think, young babies are seen as close to "culture free" individuals as possible. More importantly, however, young children are "ideal" models of language learners because they acquire their first languages as "native speakers," presumed to speak with no (foreign) accent, and no grammatical errors compared to other speakers of their dialects. Because the model of first language acquisition is considered both ideal and knowable, prospective language teachers are encouraged to study theories of "normal" language acquisition among children, and, often, to develop pedagogies that mimic this "normal" process when teaching a second language in the classroom.

As Gardner (1985) writes, although cognitive scientists are aware that human beings always belong to some culture, and always have feelings, they have designed their professional methodologies to "filter out" these variables and see human beings in their natural state as they were designed for learning and cognition.

This chapter is written as an exploration of the concept of the "natural" child created by these theories and of the suitability of this model for developing second and foreign language pedagogies. It explores questions such as the following: "Is it possible to create a politically neutral, culture free understanding of how children develop and acquire language?" "What dangers are involved in creating a single image of the 'natural child' who learns to speak, and explores the world through verbal imagery?" "Can this model of 'natural' child language development be appropriately applied to older children and adults learning second and foreign languages?" "Do our assumptions about language learning based on 'normal' models of child development lead to comfortable language learning environments for all, or do these models feel comfortable and 'natural' for some, while requiring others, in the words of Johnny Clegg and Savuka, to put on stiff and unnatural costumes, and to 'walk in the dreams of the foreigner?' "

In this chapter, I use the fields of language development and language teaching to question the use of universalized models of the "natural child" as templates for "normal" models of learning and for curriculum development. Instead, I argue that our understandings of basic human cognition in small children, which may appear "precultural" and "universal," are nonetheless shaped and understood by scientists in ways that are historically contingent and culturally loaded.

I make my arguments in the following way. First, I look at understandings of early childhood language development as a scientific metaphor. That is, I look at changing metaphorical understandings of the "natural" child who is learning or acquiring his first language. I then look at "homologies," (Popkewitz, 1998) or noncausal similarities between our understandings of child language acquisition and understandings of the "modern" productive individual drawn from other fields of inquiry. In this case, I compare our models of the "natural" child language learner to the account provided by Giddens (1991) of the culturally and historically specific entity of the "late modern" individual, to the construction of the "modern" citizen necessary for economic productivity posited by "development" literature, and understandings of the "successful" and "entrepreneurial" worker which emerges in the language of the "new" capitalism. Finally, I return to the field of language pedagogy and argue that our understandings of the "good" or "natural" child in language acquisition inevitably shape our concept of the "good" second and foreign language learner in a way that is unequally accessible to people from different cultures.

Science and Reality

Science is about truth. However, at the same time, it is also always about culture, and about politics. Latour (1993) refers to the inextricable linkage of science, culture, and politics as a "Gordian knot."

> Scientific facts are . . . constructed but they cannot be reduced to the social dimension because this dimension is populated by objects mobilized to construct it. These objects are real but they look so much like social actors that they cannot be reduced to the reality "out there" invented by the philosophers of science (p. 6).

That is, truth, culture, and politics are so closely tied together in our scientific theories, that they cannot be easily separated from one other. On one level, scientific theory appears neutral and transparent because it is demonstrably "true," and is backed by observation and by testing. However, scientific theories are also not separable from the cultural beliefs and assumptions of the researchers who develop them, the gatekeepers who selectively disseminate them, and the "consumers" who read and believe in them. Furthermore, science always grows out of, and results in, power relationships. There are clear power relations selecting whose truths are valid and whose words are heard in the scientific community, as well as how theories must be packaged to appear valid in scientific discourse. In addition, as we see to be the case with models of child language acquisition, scientific theories always have political implications and consequences.

Like Latour, I avoid implications that models of "natural" child development are not "true" or that the observations they are based on are not scientific or replicable. I believe that there are certain species specific characteristics that make beings identifiably human and I agree that carefully constructed scientific research gives us a picture of humans that is more than random or arbitrary. At the same time as I respect the integrity of scientific research, however, I also argue that our understandings of child development are always cultural as well as truth-based and always have power implications. Attempts to view understandings of the young child as a language learner as universal and neutral establish a multiplicity of ways in which all of us are "abnormal" and "deficient" because we do not match the universal model (see Rose, 1990). Furthermore, because these theories arise in specific cultural and historical contexts, they match people from some backgrounds more than others, making a hierarchy of normalcy.

Who is the "Natural" Child in the Context of Language Acquisition Theory?

Our current understandings of how children learn language are grounded in a language that "makes sense" to contemporary readers so that they seem self-evident and natural. In fact, our current images of the young child as a language learner arose relatively recently and were in distinct contrast to the models and theories which were previously popular. One can say that our current understandings of the "natural" process by which children learn language represent a fairly dramatic scientific and metaphorical shift away from the understandings that preceded them.

For much of the first half of the century, at least in the context of the United States,[1] the fields of linguistics and the psychology of cognition, or as it was called at the time, language learning, were dominated by the schools of behaviorism in psychology, and by what de Beaugrande (1991) calls "physicalist" descriptive linguistics.[2] Both of these tended to privilege directly observable facts and external behaviors and influences over "mentalist" or internally generated factors in language learning and language use. These tendencies represented a reaction against mentalist or "metaphysical" (Bloomfield, 1933) models of human cognition prevalent in the late nineteenth and early twentieth centuries, which theorized extensively about factors that could not be observed or tested.

In physicalist, or behaviorist theories, linguistics and psychology are made to resemble as closely as possible the natural, or physical and biological sciences, of the first half of the twentieth century.[3] As John Watson (1913), often described as "the father of behaviorism," puts it: "Psychology as the behaviorist views it, is a purely objective, experimental branch of natural science . . . Introspection plays no essential part of its methods" (p. 158). Within this school of thought, theories and hypotheses about how people learn and use language should be based on observation and should be provable and replicable. Unfortunately for both linguists and psychologists, many of the components that make up both speech and language acquisition are not directly observable. For example, Leonard Bloomfield (1933) laments in his classic book *Language* that "the working of the nervous system . . . is not accessible to observation from without" (p. 33). Because the inner workings of the human brain were not directly observable during much of the period of behaviorist ascendancy, linguists and behaviorist psychologists of the era assumed that the workings of the unobservable inside, were best inferred from what *was* observable—external stimuli and cause and effect chains. Bloomfield uses the following simple story to explain how children begin to speak through stimulus and response. By getting a desired reaction to something they say, children learn that speech can be rewarding. They have incentive to develop a collection of useful speech habits.

> Suppose that Jack and Jill are walking down a lane. Jill is hungry. She sees an apple in a tree and makes a noise with her larynx, tongue and lips. Jack vaults the fence, climbs the tree, takes the apple, brings it to Jill and places it in her hand. Jill eats the apple (p. 22). The normal human being is interested only in S and R [stimulus and response]; though he uses speech and thrives by it, he pays no attention to

it . . . It, along with the rest of speech, is only a way of getting one's fellow-men to help (p. 26).

Using a process similar to that in the story, children are assumed, within this framework of understanding, to learn language on the basis of stimulus and response, building up collections of tiny speech habits which are direct imitations of speech behaviors they have heard but in new and unexpected combinations. Although scholars in this school acknowledge that there is a basic human inclination to produce speech behaviors, the production of speech sounds is seen as, in itself, useless, unless they are shaped and rewarded from outside. Because the incentives, as well as the habits, for speech come from outside stimuli, the child plays a relatively passive role in this account of language learning. For these scholars, language learning is "a type of learning that involves the establishment of a set of habits that are both neural and muscular, and that must be so well learned that they function automatically" (Brooks, 1960, p. 21). In order to learn to speak, a child "makes mouth movements repeating" (Bloomfield, 1933, p. 30) sounds he or she hears from parents or other fluent speakers and forms a new habit. Speech requires little or no inner drive or creativity. "To the end of his life the speaker keeps doing the very things which make up infantile language learning" (Bloomfield, 1933, p. 46). Child language learning is thus a simple and outer-directed process. It only sounds complicated and creative because children are subjected to complex and unpredictable stimuli. Metaphorically, children can be seen as empty vessels waiting to be filled with habits through the application of external stimuli.

This understanding of the "natural" process of language learning among children is linked theoretically with a particular style of language teaching, both in second and foreign language classrooms and in programs such as Head Start (which was initiated within a behaviorist frame of arguments). A related, though updated, type of theory concerning the teaching of "known knowledge" underlies a number of curriculum methodologies popular to this day as exemplified by the underlying ideas that lead to standardized testing—that is, it is possible to both control and measure young people's learning from the outside. In this philosophy of learning, children are assumed to be in need of a particular, knowable set of speech skills. Children who do well in school-based learning, and, by extension, in later life, are assumed to have "more" of this type of knowledge, while students who are "disadvantaged" or "at risk" are assumed to have less. In addition, children are assumed to be in need of extensive structure and guidance from parents, teachers, and curriculum designers if they are to develop the correct habits and knowledge sets necessary to succeed in life.

This understanding of child learning and language acquisition was gradually replaced, in the fields of theoretical linguistics and psychology beginning in the early 1940s, although the new theories and new metaphors for child language acquisition did not begin to seriously impact related fields such as applied linguistics (the science of language teaching), early childhood education, literacy studies, and public policy until about twenty years later.[4]

Our current understandings of cognition and language acquisition among young children are the fruits of a period of intense reevaluation and reaction against such outer-directed ways of explaining human learning and behavior. This reevaluation began in the mid 1950s, but did not become prevalent in educational discourse until a quarter of a century later. This reorganization of the way we understand human cognition and learning substituted a "mentalist" (Gardner, 1985) or interior driven theory of human speech and of child language acquisition for the then dominant externally driven or behaviorist one.

First, such scholars—the budding cognitive psychologists and generative linguists—questioned the concept of complex behaviors such as speech as linear chains of associations, in which each word acts as a stimulus for the next. Using an example borrowed from Hunt (1994, p. 277), if we try to remember a multi-digit number, such as our phone number, we first remember the first number which acts as a stimulus for the second, which in turn stimulates the third, in a chain of linear associations. Linguists like Chomsky (1959) and psychologists like Karl Lashley (see Jefress, 1951) began to assert that linear stimulus response theory cannot account for behavior as complex as human speech. Instead, linguists and psychologists began to assert that language is processed hierarchically (Chomsky, 1957).

In addition, Chomsky, and other generative linguists began to insist that people have a set of internalized intuitions about language, which are not derived from any external stimuli they have encountered. The internally driven nature of these intuitions is demonstrated by the fact that we can recognize that certain sentences such as the now famous "Colorless green ideas sleep furiously" are grammatically correct, even though they make no sense in terms of real world meaning (see Chomsky, 1957, 1965). For Chomsky, such intuitions are proof that grammar is an internalized and instinctive human capacity, which exists to some degree independently of learning. Such intuitions cannot be derived from real world stimuli habits or other types of external learning because they refer to situations that are highly unlikely to be encountered outside the human mind. In fact, so interested were the new generation of linguists and psychologists in the idea of intuition that they began a series of intensive investigations of their own interior language intuitions with no empirical testing whatsoever.

Much of the focus in this new line of theoretical speculation was on the young child and the way that he/she first began learning a language. In order for a Chomskian perspective to work—for people to develop such complex inner representations of language—young children must be born with templates for acquiring language already in place. If language cannot be completely learned from without, babies and young children must possess some sort of language instinct, which guides them in understanding and creating language structures never before produced or heard. At various times in his career, Chomsky has referred to this as a "language acquisition device" or "universal grammar" (UG).

> [The child is born with an active, theory creating and testing way of learning language, called the language acquisition device. The child creates and tests] a class of possible hypotheses about language structure, [and] determin[es] what each implies for each sentence, [so he or she can] select one of the infinitely many hypotheses compatible with the given data. (1965, 5, p. 30)

This same inborn capacity for language is what Pinker (1994), in a more popular medium, refers to as the "language instinct." The idea that children start with an instinctive understanding of what language is, and orchestrate their own language acquisition process is now widespread, and is dominant to the point of being virtually unchallenged in fields such as linguistics and language teaching. Many linguists and cognitive psychologists began to view young children as cybernetic devices, already programmed with software to search out and classify whatever "language data" they encounter (Minsky, 1963; Newell *et al.*, 1963; Simon, 1969).

These works and a multitude of similar ones represent more than a theoretical refinement. They represent a fundamental metaphorical shift in the way that social scientists understand both human cognition and human motivation. Previously, the human brain was regarded as a relatively unstructured container, into which society placed language information. Children learned language because more skilled speakers, through a process of modeling and reinforcement, placed information in them. After this metaphorical shift, the brain was described as a computer, referred to in earlier works as a "cyber mechanism," or an internally driven search mechanism. Furthermore, as a cyber mechanism, the child's brain has already been programmed at birth with the goal of learning language and with complex strategies in place for reaching out and finding whatever data it needs for building a grammar and vocabulary for whatever language(s) it encounters. Older, more fluent speakers who are teaching a young child to speak play a very different, and less active, role than in the older model. They are now reduced, in strong cognitive theories[5] to two roles. The first, is to activate the child's mechanism by providing linguistic stimulus. This is like clicking the install button on a piece of self-activating software. The second role is to supply language data, which is typically viewed as incomplete and corrupt, for the child's mind to classify and arrange.

The natural child, in this model, is an active learner, who is innately motivated to reach out, grasp, and process information in a sophisticated way—forming hypotheses and seeking out necessary information in an active manner. Parents, teachers, and the external environment play a secondary role for children who are envisioned as self-motivated, driven for knowledge, and possessing an interior capacity for producing and testing hypotheses, making them capable of forming hypotheses from even incomplete and corrupt data.

Second Language Teaching Based on This Model

This model of the "natural" process of language acquisition in young children has had a dramatic impact on the fields of language learning and language teaching. Young children have a reputation of being "ideal learners," in that they tend to learn first and second languages with no grammatical errors and no foreign accent. Thus "learning like a young child," has become the gold standard in language teaching that most curricular and methodological theories try to imitate.[6] Many contemporary theories and methodologies of second and foreign language teaching follow a similar model. They are predicated on the idea that a "good" learner functions like a "natural child" learning his/her first language.

Because children, when undergoing a "natural" process of language learning (see Terrell, 1981 for a description of how to model language learning among older children and adults after the "natural" process used by young children), are assumed to be active and self-motivated learners, processing language information in a subconscious but purposeful way, many theorists now assume that second language learners can and should do the same. When developing methodologies for teaching second languages, many contemporary theorists model their approach on the "ideal learner," that is, the "natural" model of the infant and young child. The appeal of this type of model can be seen clearly in the following quote from the well-known advocate of "communicative" language teaching approaches—Stephen Krashen.

> According to [my] hypothesis, second language learners have two distinct ways of developing ability in second languages. Language acquisition is similar to the way children develop first language competence. Language learning is different [and much less effective.] It is knowing about language or formal knowledge about a language. Acquisition is far more important. It is responsible for our fluency in a second language, our ability to use it easily and comfortably. Conscious learning is not at all responsible for our fluency. (1981, pp. 56–57)

The role of the teacher also changes radically in this new understanding. Because learners are expected to be active, and to take responsibility for their own learning, teachers no longer play an active role in language teaching. Instead, their role becomes one of providing comprehensible and accessible data for students to process with their active learning capacities. Again citing from Krashen's (1981) classic account of "natural" language learning and teaching, we can see the strong influence of the active, cybernetic model of the child language learner.

> The best way to "teach" speaking . . . is simply to provide "comprehensible input." Speech will come when the acquirer feels ready. This

readiness state will come at a different time for different people. (Krashen, 1981, p. 59)

Is This Model Truth, Culture or Power?

Natural Learning and the New World Citizen

Fifty years of work in cognitive psychology, in linguistics, and in child language studies have shown many ways in which the new cognitive models explain child language acquisition better than older theories. Furthermore, child language acquisition theorists and cognitive scientists now have years of empirical research on which to base their originally fairly theoretical assertions about how children learn language. (See, for example, the wide range of empirical studies included in Bloom, 1996.) It would be difficult at this point to argue that there is not a strong "truth" element in cognitively oriented theories of child language acquisition. However, are these theories more than empirically testable "truths?" Is there also a cultural and a power element to them?

Implicit in our models of young children first exploring the world and their place in it are concepts of children as protoadults and protocitizens. We understand from the fields of anthropology and comparative linguistics that the child is born with a certain amount of cognitive and cultural flexibility. This enables him/her to learn any of the world's languages and to participate in any of the world's cultures. At the same time, scholars like Chomsky assert that there is a common, "universal" structure underlying all human languages, while others, like Gardner (1987) assert that it is possible to strip away factors like "affective factors or emotions," and "the contribution of historical and cultural factors" to come up with a universally applicable model of the human being. Is it truly possible to invent a model of "natural" child development and cognition that has been stripped of the cultural, the historical, and the political?

To explore these aspects of our current models of the "natural" child acquiring language, I look at the currently popular model in several ways. First, I look at "homologies" between the metaphorical structure of these theories and other concepts of the late twentieth century individual and "productive" citizen. Second, I question whether these theories provide an equal playing field for students from all cultural and social backgrounds.

The "Natural Child" As a Model of the

Individual and the Productive Citizen

By this point readers will be aware that contemporary concepts of the process of language acquisition among infants and small children are more than statements about how babies learn to speak. They are also metaphorical ways of conceiving of the individual and his or her way of fitting into the world. We can also see that the metaphor underlying and shaping our understanding of the "natural" child has changed fairly dramatically over the last five decades from that of a relatively passive, outward-directed being who requires information, structure, and motivation from outside to that of an inner-directed, self-motivating individual who actively seeks out and processes information.

Not surprisingly, the same type of concept of the human individual emerges in other arenas besides language learning and teaching. While we can say that there is some degree of causality in the relationship between first and second language acquisition theory, the relationships between the child constructed in language acquisition and the productive citizen in others is better described as homologous (Popkewitz, 1998)—that is, both emerge from more general understandings without direct causality. In this section of the chapter, I look at the relationship between our contemporary models of the "natural child" language learner and other conceptions of the "late modern" individual and of the post-Fordist productive worker.

The "Natural Child" As a "Late Modern" Individual

In this section, I look at a concept of human individuality which is intentionally historically and culturally specific—that of Giddens' (1991) model of the "late modern" individual. In this account of what it means to be human in the context of a "late modern," or "high capitalist" context. This model is intended to describe a specific way of being developed at a particular time, and in particular and unique cultural and historical circumstances, and it is explicitly contrasted with other contexts in which the individual is seen in a different way. What is striking about Giddens' account is how closely it resembles the metaphorical understandings of the "natural" language learner put forth by people like Chomsky and Krashen.

For Giddens, the "late modern" individual is striking in the degree to which he/she is required to shape knowledge and identity *within the self* rather than learn from, and conform to, tradition. Each step in life requires an intense, internalized set of decisions and evaluations created within as much as from outside influence.

> In the post-traditional order of modernity, and against the backdrop of new forms of mediated experience, self-identity becomes a reflexively organized endeavour. The reflexive project of the self—which consists in the sustaining of coherent yet continuously revised biographical narratives, takes place in the context of multiple choice filtered through abstract systems (p. 5).

This process of continuous self-construction—the building of a theory of who one is in the world, results in the development of an individual with particular characteristics. Below are some of the characteristics of the "late modern" individual as described by Giddens (1991).

1.   The self is seen as a reflexive project for which the individual is responsible. We are what we make of ourselves.
2.   The self forms a trajectory of development from the past to the anticipated future. The individual appropriates his past by sifting through it in the light of what is anticipated. The trajectory of the self has a coherence that derives from a cognitive awareness of the various phases of the lifespan. The lifespan, rather than events in the outside world becomes the dominant "foreground figure" in the Gestalt sense.
3.   The reflexivity of the self is continuous as well as all–pervasive. At each moment, or at least at regular intervals, the individual is asked to conduct a self-interrogation. Beginning as a series of consciously asked questions, the individual becomes accustomed to asking: reflexivity in this sense belongs to the reflexive history of modernity as distinct from the more generic reflexive monitoring of action (1991, pp. 75–76).

Obviously the young child as described in theories of linguistic development is not consciously aware of the "lifespan," nor is he or she consciously self-reflective. However, the same mechanisms of goal seeking, evaluation, and self-motivated trajectories play a striking role in both concepts. Just as babies and young children take information given to them by their cultures—that is, language input or language data—evaluating it and arranging it to make a coherent theory of who they are as speaking beings, the adult citizen in a "late modern" society takes in information and processes it internally, working continuously *on the inside* without external input, shaping or motivation to create his/her own model of the self. Similar to the language young children reach out, process, and assemble, a language that others can understand and respond to, a selfhood that fits into a particular social model is reflexively constructed by Giddens' "late modern" individual. If this were not the case, it would not be possible to describe "late modern" culture, or the "late modern" individual in a coherent way or to contrast this individual with the model of the self in other times and places. However, what is distinctive about Giddens' individual is the fact that the work of constructing who one is in the world seems to come from within.

The Self-motivated Theory Constructing Child
and the Entrepreneurial Worker

If the interiorized, theory building child is like Giddens' model of the individual, he or she also looks strikingly similar to the new model of the "productive worker," put forth in the literature of what Gee *et al.* (1996) call the world of "fast capital" and I call the world of the "cybernetic worker." A concept of the self-reflexive, self-motivated, interiorized individual who resembles Giddens' "late modern" individual, as well as Chomsky and Krashen's active, self-motivated theory building child, emerges in business literature. As Gee *et al.* (1996) point out, enormous economic and structural changes in industry have taken place in the last 50 years.

> The new capitalism as defined in the new capitalist literature is not about commodities or standardization, and very probably not about democracy. The new capitalism is . . . about customization: the design of goods and services perfectly dovetailed to the needs, desires and identities of individuals on the basis of their differences . . . Thus the new capitalism celebrates diversity and abhors standardization. The new capitalism is not about democratizing desire, but rather about customizing desire (p. 43).

These changes have also resulted in the emergence of a new model of the worker, who I term the new "cyber" worker. Along with the changes in industry, a new type of worker with an interiorized, success and knowledge seeking "servo-mechanism" has been birthed. The self-help author Maxwell Maltz (1960) described this newly emerging "cyber-worker" over forty years ago, in a manual of advice concerning how people could become more professionally successful in a rapidly changing and more competitive corporate world. In this new concept, individuals should envision their minds as "guided missiles," using information and feedback to fly towards ever-changing and ever-increasing goals.

> The new science of "cybernetics" has furnished us with convincing proof that the so-called "subconscious mind" is not a mind at all, but a mechanism—a goal-striving, "servo-mechanism" consisting of the brain and nervous system . . . This automatic, goal striving machine functions very similar [*sic*] to the way that electronic servo-mechanisms function, as far as basic principles are concerned, but it is much more marvelous and more complex than any electronic brain or guided missile ever conceived by man (p. 12).

A more contemporary example of this way of thinking comes from the job search guru Jeff Taylor, founder of the internet employment agency, *Monster Careers*. In this model, every worker is a potential job seeker and every job seeker is an entrepreneur, taking on the cybernetic role of managing his/her career by seeking out job opportunities in a self-motivated, self-guided way.

WORK LIKE AN ENTREPRENEUR.
Are you willing to pour all your energy, imagination and discipline into finding your dream job? Then you're a lot like entrepreneurs who start their own businesses. Here's my definition of an entrepreneur: when everybody around you thinks you're crazy and you still

think you have a great idea—and you will put yourself on the line for that idea. The work habits of entrepreneurs can teach you a lot about how to run a successful job search. They never quit. Failure is not an option for entrepreneurs. They learn from their mistakes and act immediately to correct them. When you have setbacks in your job search—and everybody does—you must treat them as opportunities to learn and get better. Entrepreneurs by definition are individualists. (Taylor & Hardy, 2004)

Not surprisingly, this new "cyber-worker" bears a strong resemblance to the "cyber-child," or the language learning child as a computer, described in the linguistics and psychology literature we have just looked at. Again, infants and young children are incapable of "programming" themselves for success in a workplace they cannot yet understand. However, the interiorized, self-motivated, active and theory building child described in the literature of contemporary linguistics and cognitive psychology looks much like the self-motivating, self-directed, and self-educating "cyber-worker" described in the passages above. Just like the "cyber-worker" works in an interiorized and self-directed way to seek success, the young child works in an interiorized and self-directed way to "reach out" for mastery of language. Just as the "cyber-worker" motivates himself/herself to have "psychological ownership" of the job, the new "cyber-child" has ownership over the language acquisition process and needs no motivating, shaping, or rewards from outside.

How Does This Contrast With Past
Understandings of the Worker?

To strengthen the linkage between the new "cyber-worker" and the "cyber-child" acquiring language, I will quote briefly from a previous era in management philosophy and compare it to early, Bloomfieldian models of how children learn to speak a first language.

In 1959, toward the end of the Fordist industrial era, Robert Saltonstall, describes how managers should motivate their workers. In this model it is clear that motivation comes from without. Employees (and I use this term purposefully, in contrast to the "entrepreneurial" worker of contemporary management literature, are assumed to be willing to work hard providing they are in a workplace that provides clear direction, clear rewards for good behavior, and sanctions for undesired actions. Like the language-learning child in a Bloomfieldian model, these workers respond primarily to outside stimuli and structures. Their productivity is shaped by providing rewards when they behave in a desired manner and sanctions, or punishment if they behave in an undesirable manner.

Our goal should be to provide a well-organized working environment where physical and mental obstacles to production are removed and where people are challenged to optimum effort because they see this as worthwhile . . . for them. This implies aggressive and inspired and sensitive leadership, high standards of performance and adequate discipline which leads to mutual respect. In such an environment employees see management creating opportunities for them to grow and utilize their physical and mental skills in meaningful work under good supervision (p. 177).

Is the "Natural Child" Natural? Can the Language Learning Process Be Described in a Transparent and Culture-free Way?

In the two previous examples, I have argued that although there is definitely "truth" in cognitive and theoretical linguistic understandings of the way children learn language, there is also a strong resemblance between our concept of the "natural" language learning child, and our historically and culturally specific understandings of the "late modern" individual and the "new" industrial worker. I have also argued that there is some degree of synchronization between the emergence of the new model of the theory creating "cyber-child" in the newly emerging generative linguistics and cognitive psychology of the 1950s and 1960s and the new "cyber-worker," who was beginning to emerge to inhabit the post-Fordist workplace. If it is possible to produce a coherent and empirically testable model of child language learning from a cognitive perspective, it may be more difficult to achieve Gardner's goal of stripping away culture and history from our concepts of what it means to be human, to produce a universal model. While there is probably something innately human, about how people learn and use language, it becomes difficult, if not impossible to determine what part of our theories are describing human universals and what parts are describing our own historical and cultural values about desirable human traits. This problem is complicated by the fact that the discursive space of the human sciences is largely constructed and inhabited by people who share similar cultural characteristics. As Chakrabarty (2000) points out, the models of what it is to be human coming from social science and philosophy suffer because they all come from a single, particularistic intellectual tradition.

> For generations now, philosophers and thinkers who shape the nature of social science have produced theories that embrace the entirety of humanity. As we well know, these statements have been produced in relative and sometimes absolute ignorance of the majority of humankind—that is those living in non-Western cultures.

By saying that all of the human sciences come from a Western cultural tradition, scholars like Chakrabarty are not implying that all of the human scientists come from Europe or North America. Instead, he is arguing, as does Pennycook (1994) that Ph.D. scholars all end up occupying a similar cultural and intellectual space, even if they do not begin there. The process of formal education as it currently exists may not be merely a process of acquiring facts. Instead, it is a process of learning how to think, to feel, and to experience one's self as an individual.

Models of "Natural" Child Learning and Power

It is in comparing Western and non-Western (and perhaps upper-middle and upper class) conceptions of learning that we come up against the concepts of models of child development, curriculum, and power. As we have seen, using the field of language teaching as an example, particular models of what it is to be a "natural" child and "naturalistic" models of teaching and learning, lead to certain models of curriculum development, while making others less likely. It is from this standpoint that I would like to briefly touch on the concept of child development and power.

Up to this point, I have argued that there is a striking degree of consonance between our current concepts of infant and child development, learning, and a culturally and historically specific understanding of the individual and the worker. At the current time, we have a model of the "natural," or "universal" child learning and developing language skills which is consonant with the concept of the individual in certain sectors of Westernized, postindustrial, late modern society. My question is, if this model of the active, language-acquiring child is truly universal and transparent, is it equally accessible to children from all backgrounds?

For a number of years, linguistic anthropologists have argued that people from various cultures have ways of participating discursively in conversations that create very different interpersonal interactions than people from "Western" backgrounds are used to (see, for example, Phillips, 1972; Scollon & Scollon, 1981, 1995). They have also argued that differences in communicative styles often cause problems in the classroom. What may be embedded in some of these differences is a concept of who the individual is and how he or she relates to the rest of society. Chakrabarty (2000) gives an example of this when he talks about the modern, middle-class individual in India who often has two quite different internal ways of experiencing himself/herself. On one level, and in some situations, modern, middle-class Indians have taken on the late modern, Western concept of the self-creating individual. At the same time, the same people have a quite different concept of their identity, as embedded in a group and family structure, without which they would not be completely whole. A similar sense of dual identity is evoked in the recent Indian novel *Bombay Time* (Umrigar, 2001) which tells the story of a group of middle-aged, "modern" couples in Bombay, who have chosen their own life paths, and have rejected arranged marriage to marry for love, but who discover as they get older, that neither their marriages, nor their sense of self, cohere, unless they are embedded in a community.

Second language classrooms, in particular, may encounter this type of cultural dissonance. Inoway-Ronnie (1998) describes conflicts concerning "appropriate" ways to teach English as a second language to Hmong (Southeast Asian refugee) children, with the parents perceiving, among other things, that "communicative," unstructured environments in which children construct their own knowledge about English, being in conflict with community-oriented, hierarchical values in Hmong society. Hmong people who have learned English and survived and succeeded in the U.S. educational system often talk about having two selves, which they do not know how to reconcile. The following quote from a young Hmong man, who has gone to college and become an architect illustrates this conflict.

> Upon graduation from my Master's program, I wanted to travel to Europe and become a successful international architect. But as a responsible Hmong son, I could not wander far from my ailing parents who had risked everything to [bring me to the United States.] My parents needed me to stay to take care of them, and to ensure the future of the clan . . . However, I was devastated when I heard that a graduate from my school was offered employment at Foster and Partners, an international award-winning architecture firm in London that I had wanted to work for as a young designer. (Cha, 2002, p. 23)

Throughout his essay, Cha discusses the painful sense of dissonance he feels as a result of having two totally different individuals inside him—the one who is closely embedded in family and clan and whose first responsibilities are to others and to tradition, and the self-reflective, self-constructed, and constructing Western individual who has controlled his own intellectual development and wants to control his life. One must ask, in cases like this, are our developmental and curricular understandings of the self-motivating and interior-directed language-creating individual truly universal and truly without cultural or political implications?

Conclusions

The point behind this piece is not to argue that our current notions of child development and related curricula and methodologies are untrue, that they are entirely cultural constructs, or that they are totally embedded in power relations. Furthermore, I do not wish to argue that the communicatively oriented language teaching programs that grow out of these conceptions of the child are "unfair" and "exclusionary" and that they should not be used in diverse classrooms. Just as it is difficult to come up with one, culture-free model of human development, it is also impossible to come up with an easy formula for teaching children from non-Western backgrounds. In fact the sense of a dual self, reported by people who have gone through "Western" educational systems but come from cultures that see the individual in ways very different from the late modern/postmodern/postindustrial model of the "cyber-individual" may be a survival technique—necessary and useful for coping with a globalized economic and educational system that demands a particular sense of self and a culture of origin that sees the individual as embedded in and shaped by collective identity.

What I do want to argue against, however, is the assumption that science alone, without the moderating hand of cultural and historical understandings, can put forth one, culturally universal way of seeing the developing child, and that school children can be judged and normed according to a singular model. Child development and language learning are partly about science and partly about truth. They may also be partly about cultural and historical specificity, and about power and exclusion. Accepting the one aspect of development models without examining their flip side—that of judging and norming those who are culturally different—is as dangerous as rejecting them altogether.

Notes

1. Early to mid-twentieth century linguistics differed in significant ways between Europe and the United States.
2. This same school is often called American structuralism. I avoid this name as it sounds too similar to European structuralism, which uses a mentalist, or interiorized model of language. The similarity in the names of two such different models can be confusing.
3. In the interim, many of the "natural" sciences have changed as well, with scientists such as theoretical physicists making theories about forces that are not directly observable.
4. For example, well-known scholars (like Nelson Brooks, who wrote a strictly behaviorist book on language learning theory for teachers, in 1960, producing a second edition in 1964,) were able to talk about first and second language acquisition from a behaviorist perspective in a way that completely ignored the work of Chomsky and others until some time in the mid-1970s.
5. Some linguists now accept a somewhat weaker version of this theory.
6. There are a few dissenters, like Strozer (1996) who, while ascribing to an understanding of language acquisition in young children very similar to that of Chomsky and others, believe that the capacity to learn "actively" disappears almost totally with age and that adolescent and adult learners must learn languages through structured instruction.

References

Bloom, P. (1996). *Language acquisition: Core readings*. Cambridge, MA: MIT Press.
Bloomfield, L. (1933). *Language*. Chicago, IL: University of Chicago Press.
Brooks, N. (1960). *Language and language learning*. New York: Harcourt, Brace and World.
Cha, B. (2002). Being Hmong is not enough. In Mai Ling Moua (Ed.), *Bamboo among the oaks: Contemporary writing by Hmong Americans* (pp. 22–31). St. Paul, MN: Minnesota Historical Society.
Chakrabarty, D. (2000). *Provincializing Europe: Postcolonial thought and historical difference*. Princeton, NJ: Princeton University Press.
Chomsky, N. (1957). *Syntactic structures*. The Hague: Mouton.
Chomsky, N. (1959). Review of Verbal Behavior by B.F. Skinner. *Language, 35*, 28–58.
Chomsky. N. (1965). Aspects of the theory of syntax. Cambridge, MA: MIT Press.
de Beaugrande, R. (1991). *Linguistic theory: The discourse of fundamental works*. New York: Longman.
Gardner, H. (1987). *The Mind's New Science: A History of the Cognitive Revolution*. New York: Basic Books.

Gee, J., Hull, G., & Lankshear, C. (1996). *The new work order: Behind the language of the new capitalism.* Boulder, CO: Westview Press.

Giddens, A. (1991). *Modernity and self-identity: Self and society in the late modern age.* Stanford, CA: Stanford University Press.

Hunt, M. (1994). *The story of psychology.* New York: Anchor Books.

Inoway-Ronnie, E. (1998). High/Scope in Head Start Programs serving Southeast Asian immigrant and refugee children and their families. In J. Ellsworth & L. Ames (Eds.), *Critical perspectives on project Head Start.* Albany, NY: State University of New York Press.

Jefress, L. (1951). *Cerebral mechanisms in behavior: The Hixon Symposium.* New York: John Wiley.

Krashen, S. (1981). Bilingual education and second language acquisition theory. In *Bilingual education: Theory and practice* (pp. 51–82), California State Department of Education. Los Angeles Evaluation, Dissemination and Assessment Center. Los Angeles: California State University Press.

Latour, B. (1983). *We have never been modern.* Cambridge, MA: Harvard University Press.

Maltz, M. (1960). *Psycho-cybernetics: How to get what you want out of life.* Englewood Cliffs, NJ: Prentice Hall.

Minsky, M. (1963). Steps towards artificial intelligence. In E. Feigenbaum & J. Feldman (Eds.), *Computers and thought* (pp. 406–450). New York: McGraw Hill.

Newell, A., Shaw, C. & Simon, H. (1963). Elements of a theory of human problem-solving. *Psychological Review* 65, 151–66.

Pennycook, A. (1994) *The cultural politics of English as an international language.* New York: Longman.

Phillips, S. (1972). Participant structures and communicative competence: Warm Springs children in community and classroom. In C. Cazden (Ed.), *Functions of language in the classroom.* (pp. 370–394) New York: Teachers College Press.

Pinker, S. (1994). *The language instinct: How the mind creates language.* New York: Harper Perennial.

Popkewitz, T. S. (1998). Dewey, Vygotsky, and the social administration of the individual: Constructivist pedagogy as systems of ideas in historical spaces. *American Educational Research Journal,* *35*(4), 535–570.

Rose, Nikolas (1990). *The shaping of the private self.* London and New York: Routledge.

Scollon, R., & Scollon, S. (1981). *Narrative literacy and face in interethnic communication.* Norwood, NJ: Ablex.

Scollon, R., & Scollon. S. (1995). *Intercultural comunication: A discourse approach.* Cambridge, MA: Blackwell.

Simon, H. (1969). *The Sciences of the Artificial.* Cambridge, MA: MIT Press.

Strozer, J. (1996). *Language acquisition after puberty.* Washington, DC: Georgetown University Press.

Taylor, J. & Hardy, D. (2004). *Monster careers: How to land the job of your life.* New York: Penguin Books.

Terell, T. (1981). The natural approach in bilingual education. In *Bilingual education: Theory and practice* (pp. 51–82), California State Department of Education. Los Angeles Evaluation, Dissemination and Assessment Center. Los Angeles: California State University.

Umrigar, T. (2001). *Bombay Time.* New York: Picador.

Watson, J. (1913). Psychology as the Behaviorist Views It. *Psychological Review 20,* 158–177.

CHAPTER 5

The Quest for Health in Different Timespaces★

Dar Weyenberg

Today's health practices constitute the individual in a new way. In the early decades of the 1800s, the health of the individual emerged as a social tool for ordering the household and society. Today, the healthy individual is conceptualized as an autonomous, active, actor who is no longer connected to the social order as he was in the previous centuries. We live in a time when schools are paying increasing attention to accountability, standards, and "best practices" as the solution to poor schooling outcomes. An underlying presumption of these initiatives is that all individuals can be assessed uniformly, using the same standards for all. This presumption can be read in the recent U.S. Department of Education (2002) legislation or in national education standards. Along with these legislative initiatives, school administrators and teachers have been encouraged to agree to and put into practice universalizing standards of youth and education for active citizenry.

The idea of standards is not new, as schools have always had notions of what it is that children need to learn, however the nature of these goals or standards has changed over time.[1] Standards function as a governing practice to ensure an equal playing field for all. Educational standards do two crucial things: they simplify complex phenomena, making for legible, calculable, and careful management. Second, drawing on Hacking, Popkewitz (2004) argues that standards are productive in that they create certain kinds of individuals with particular dispositions—and not others. We can consider standards in this double manner when examining any school subject matter, including history, mathematics, or health, among others, as each have promulgated their own national and state standards. As an effect of U.S. Department of Education legislation, schools have more closely adhered to the national health education standards as the gold standard for what students should know about health (Joint Committee on National Health Education Standards, 1995). These national standards along with

TOURO COLLEGE LIBRARY

other governmental health initiatives, such as the Center of Disease Control and Prevention (CDC) focus on a relation of medical, physical, psychological, and moral categories of health. For example, the CDC has identified six categories of risky behaviors that are inserted into the national health standards; behaviors that contribute to intentional and unintentional injuries, tobacco use, alcohol and other drug use, sexual behaviors contributing to pregnancy and sexually transmitted diseases, dietary patterns contributing to disease and insufficient physical activity (Meeks *et al.*, 2003, p. 4).

These standards are brought into health textbooks and translated into specific curriculum health topics. Health texts are important not only for the standards that they enunciate but also for the standards and rules through which they construct the notion of the child (and society) as healthy: morally, ethically, and physically.

The textbooks, for example, have considerable capacity to shape the direction of the health curriculum in that they provide the all-important foundation or scientifically derived rationales for particular health related subject matter. To establish health education with a scientific knowledge base, textbooks are anchored in a matrix of sociological, medical, physiological, psychological, public health research studies, and U.S. government reports and initiatives that subscribe to certain contemporary understandings of youth and health. These textbooks also incorporate the recommendations of significant policy documents from health professional organizations such as, among others, the American Medical Association, the American School Health Association, and the Society for Public Health Education.

Today's health standards construct a particular human kind—the "health literate" child. This child is comprised of four essential characteristics. This child is (1) "a critical thinker and problem-solver" who can 'identify and creatively assess health problems and issues at multiple levels—personal, national and international; (2) "a responsible and productive citizen". . . who realizes his "obligation to ensure that their community is kept healthy, safe, and secure so that all citizens can experience a healthy quality of life. They also realize that this obligation begins with self" and avoids conduct that may harm self or others; (3) a "self-directed learner" who has "a command of the dynamic health promotion and disease prevention knowledge base" and who has the capacity to "use literacy, numerical skills, and critical thinking skills to gather, analyze, and apply health information as their needs and priorities change throughout life" and can "use interpersonal and social skills" to learn about and from others to "grow and mature toward high-health status"; and a health literate child is; (4) an "effective communicator who organizes and conveys beliefs, ideas, and information about health through oral, written, artistic, graphic and technologic mediums . . . able to convey respect and empathy towards others through the creation of "a climate of understanding and concern for others" (Joint Committee on National Health Education Standards, 1995, pp. 5–7).

To be health literate involves a double logic. That logic relates to norms of the characteristics of the child who is healthy. At the same time, the standards are written about deviance and a moral fear. The standards embody norms such as, for example, developing the capacity to avoid risky behaviors. These risky behaviors are ordered by the *National Health Education Standards*, and "performance indicators" that designate what students should know and be able to do.

Thus health is not only about being disease free, but is connected to particular notions of bodies, standards, practices of well being or norms of existence (life), and cultural norms and values, all of which engage in governing of the self and others. The healthy child of today as constructed in educational reform discourses is made to appear seamless—as if this present configuration had always been there. In addition, childhood is understood as a natural state and the healthy child is fabricated through developmental accounts such as grade levels or developmentally appropriate pedagogy—as the creator of his/her own life.

Standards and performance indicators are an important part of the governing practices of modern schools, in that they are authoritative prescriptions for thought and action. The orchestration of health disciplines, professional organizational mandates and US governmental initiatives, along with their associated techniques make the "body legible." This body is a social body as well as a biological body and has been increasingly "marked . . . from the detail of the (individual) corporeal body to the mind of everyone" (Armstrong, 1998, pp. 24–25). Once the body is made legible or readable through devices as classification systems and visual grids, administrative strategies such as health pedagogical strategies and interventions for youth can be deployed.

To consider the construction of this social body requires considering today's health literate child historically. The focus of this chapter is to problematize and unpack the notion of the healthy child in contemporary schooling practices through a history of the present. The first section considers the intellectual approach taken in this analysis, a history of the present. In the sections that follow, I look at early and mid-nineteenth century health advice: health emerging as a problem that one could improve upon; work on the self and a gendered curriculum as a strategy to achieve health. I use texts primarily written by Catharine E. Beecher as exemplars to talk about health reform advice. I do not address "medicine" per se, as in the sense of progress. Rather, I explore how medical or scientific knowledge functioned within health reform discourses to produce certain dispositions (and not others) and how these knowledges were articulated in terms of normative constructions about how individuals were to conduct their lives and maintain their health.

A Framework for Analysis: A History of the Present

As used in this chapter the term history of the present muddles the taken-for-granted naturalness of today's health literate child. Our present ways of thinking about the healthy child share an inheritance or ancestry with other timespaces. I explore the constructions of the healthy child in early nineteenth century to make visible both the breaks and continuities in how health discourses produced particular kinds of subjectivities. This sense of history is not an evolutionary, progressive sense of history; rather it brings in the local and contingent conditions that govern us in the name of health. This view is one that has echoes of the past in the present, but also encompasses breaks or ruptures in its systems of reasoning about health. Threads from early configurations mix with new and shifting forces to reconfigure today's healthy child. These shifting historical constructions are not an ending in themselves but function to govern the child, parents, and educators through health discourses.

Subjectivity is an individual's sense of self. People do not arrive in the world with a given subjectivity; it is produced through our experiences and interactions with others. Lupton (1995) notes that central to the constitution of subjectivity is discourse, and argues that subjectivity is "fragmented . . . changeable" and that "there are numerous, often contradictory ways in which individuals fashion subjectivities." This is not to say that individuals have an unconstrained choice of subjectivities, as "subjectivity is constructed through and by the articulation of power" (p. 7). For example, while there is a range of possible subjectivities available for motherhood, there are also limits consistent with broad cultural understandings and assumptions of mothers in a given timespace.

We can think of health discourses as constructing subjectivities that are effects of power. Various fields of study discursively construct and normalize what is meant by health (and therefore to be healthy or to become healthier). Fields of study such as physiology and anatomy were taking on new importance in health reform strategies of antebellum America. In addition, newer areas of scientific study, such as chemistry, were gaining status and lending justifications to health advice. Health reformers marshaled these emerging disciplinary knowledges to legitimatize certain forms of conduct, shaping particular subjectivities.

In my discussion of early nineteenth century and contemporary health discourses, I use the notions of health technologies and governmentality. Health discourses can be thought of as "the surface of emergence or the points of application for certain disciplinary or regulatory technologies" (Osborne, 1992, p. 83). I develop the term "health technologies" from Foucault's use of technologies as a term to describe practices of the self or techniques for acting upon ourselves. This work on the self creates particular dispositions and subjectivities. Health technologies, including health promotion, create desires and attachments to certain kinds of knowledges or "truths" about health. We can view technologies as the relations of knowledge and power, and therefore as effects of power.

I also use Foucault's analytic of governmentality to explore health education practices. Governmentality allows for the linking of the emergence of disciplines related to health and its knowledges and strategies that function to offer advice on how individuals should conduct themselves. To "govern . . . is to structure the possible field of action of others" (Foucault, 1983, p. 221).[2] This form of government in not centered in state or sovereign power, but governs through expert knowledge or regulation "at a distance" (Petersen, 1996, p. 49). In this sense, health education discourses are productive, in that they create particular kinds of individuals through their valorized and expert systems of knowledges. Health discourses, as systems of reasonings, normalize the way in which individuals are judged, seen, talked about, and acted upon by others and themselves as particular kinds of individuals (Channing, 1994).

A Shift from Health Education to Health Promotion

In this chapter, I use the term "health education" in reference to early and mid-nineteenth century health discourses. Contemporary health discourses are more appropriately called "health promotion." Health education and promotion discourses are governing practices that order, classify, and differentiate the characteristics of particular kinds of individuals. The discourses are not only of the school, as they traverse many institutions related to medicine, family, childrearing, and health clinics. Discursive practices regulate subjectivities by creating normalizing images of what it means to be a healthy person. These changing notions of what it means to be healthy produce different discursively constructed bodies with certain capacities. They also kindle, in different ways, a desire for health, further encouraging individuals to make themselves continuous objects of self-scrutiny and self-monitoring. Health discourses, drawing on strategies of normalization, characterize academic disciplines such as biology, public health and health education, which constantly measure, assess, record and project the risks and limits of health. Both can be thought of as forms of life-conduct, shaping ethical existence while at the same time functioning to optimize capabilities of the body (Dean, 1994). Traditionally, it was the "sick" body placed at the center of such monitoring. However, with this reasoning, even the body classified as in "good" health is also subject to governing practices such as perpetual monitoring and regulation by the self and others.

Both sets of discourses place the responsibility for good health on the individual, use professional expertise, and include prevention strategies. However, there are significant differences. Health education has a stronger focus on prevention of disease or injury. Advice could include lists of fatty foods to avoid, immunization schedules to follow, or proper seatbelt usage. Additionally, choice is limited, as the advice given by the expert is self-evidently taken to be the healthy (and only) choice. In this conception, health experts know the ideal choice and attempt to persuade the individual to follow this advice.

Health promotion, as a relatively new concept, differs in that it "focuses on empowering people to control their own health" (Gastaldo, 1997, p. 117). It directs its attention to maximizing health reserves and capacities and emphasizes community action through partnership and coalition building (e.g., Breslow, 1999). Health promotion discourse creates subjectivities that are different from earlier configurations in that they construct "customers," and "clients" who are expected actively to seek out "ways of living most likely to promote their own health" (Rose, 1999, p. 87). In this way of thinking, self-governing individuals are positioned as their own "experts," problematizing aspects of selves and their lives in the name of health, make prudent health choices drawn from a range of possibilities, and are expected to act responsibly on these decisions. Health promotion discourse is one of the new reform pedagogies that are to "empower" the individual to develop the capacity for problem-solving for present and future self-management of choice and conduct of life.

The Emergence of Health as a Problem

Health emerged as problematic in the early decades of the nineteenth century and a major shift was underway in how people understood health. People were now realizing that they could do something positive to improve their health. A shift was also occurring in the way people viewed the causes of ill health. No longer was illness attributed to chance or as punishment from Providence (Rosenberg, 1997). Health problems were now located in the actions of the individual. But before people could act to improve their health they needed knowledge. Much was self-learned through pamphlets, books, and public lectures, but this education was also to take place in the newly established common schools. Schools, as well as homes were privileged sites to instill healthful living habits. Many reform movements, such as housing, schooling, dress, and temperance coalesced around a common concern for the health of the individual and that of the nation.

Health as a Mother's Duty

The health status of women, as mothers of the nation's future citizens, was a primary concern for many health reformers (i.e., Our Daughters, 1857). European travelers to the United States focused on health issues, particularly the health of women, as it was a frequent discussion item in their letters in the early decades of the nineteenth century.[3]

Beecher envisioned educating the self and others, especially girls as future mothers, as a duty. In an essay, *Suggestions respecting improvements in education*, she argued that in order to restore and prevent health problems females needed to be educated about the functions of their bodies. For Beecher, being a woman was synonymous to being a teacher. Querying her readers, she asks:

What is the profession of a woman? Is it not to form immortal minds, and to watch, to nurse, and to rear the bodily system, so fearfully and wonderfully made, and upon the order and regulation of which the health and well-being of the mind so greatly depends? . . . have you ever devoted any time and study, in the course of your education, to any preparation for these duties?

Beecher was critical of the subject matter of schooling for girls as well as pedagogical methods such as rote memory learning. She argued that the typical curriculum for females did not teach them useful skills or knowledge. In this essay, Beecher asked her female readers if they were . . . "taught anything of the structure, the nature, and the laws of the body which [they] inhabit?" and "Have the causes which are continually operating to prevent good health and the modes by which it might be perfected and preserved ever been made the subject of any instruction?" Beecher imagined a "NO" response from her rhetorical questions she posed to her readers. Instead of useful knowledge, Beecher asserted that schooling practices taught young women more about . . . "the structure of the earth, the laws of the heavenly bodies, the habits and formation of planets . . . more of almost anything than the structure of the human frame, and the laws of health and reason" (1829, pp. 7–16).[4]

Including health knowledges in common school and female seminary curricula was an important strategy for reformers. The purpose of schooling was to instill habits and character, not merely to advance the intellectual capacities of the individual. Health habits could be instilled like any other habit.[5] The underlying reasoning was that good health was an outcome of individual habits (Bartlett, 1838; Blackwell, 1852). Highly critical of "ornamental" subject matter, such as French or drawing in schooling for young women, Beecher advanced her own health curriculum reforms and agitated for the inclusion of health-related content such as physiology and physical education in all schools. In Beecher's last reported lecture, she expressed her primary aim for schooling: "The adaptation of women's education to home life" (p. 13). A distinctive domestic curriculum was designed for school-aged girls who were constructed as future wives and mothers. In debates about the type of education appropriate for the two sexes, a common theme to emerge in the early republic was an "education for life." This meant that the curriculum should reflect the roles children would be expected to fill later in life. Since much contemporaneous thought held that major differences existed between the sexes in all areas including the intellect, the curriculum was to be designed differently for girls than for boys.

Beecher was writing in a timespace when a transition was occurring in conceptions of motherhood. This transition was an enabling condition allowing Beecher to construct motherhood as having a particular need for education. New conceptions of mothers as the primary childrearers in the family and responsible for the education of the children made Beecher's interventions thinkable. Bloch (1992) examined the literature about motherhood circulating in America between the late seventeenth and the early nineteenth centuries. This literature did not specifically address the role of women as mothers. Rather, the emphasis was on their roles as wives or "help-meets" (p. 4). Further, when women were discussed, mothering was but one of their many responsibilities within the family, as the father was primarily responsible for the education of the child. Bloch argues that a gradual shift occurred in which women were to become the primary parent in the family responsible for educating the children. The effect of this transition was that it became possible for Beecher to advocate schooling for all children, and in particular, a distinctive curriculum that would prepare girls for "the peculiar responsibilities of American women." For her, women were the primary "agents in accomplishing the greatest work that ever was committed to human responsibility . . . [the] molding and forming of young minds" (p. 39). Single women had the duty to be "mothers" in the sense that they were to take in and provide care for orphans or other "deserving" folks in their homes (1842). These discourses on motherhood produced differences or distinctions, contributing to the "separate spheres" rhetoric or that nature (or God) constituted men and women differently. Embedded in her texts was the necessity of developing the habit of placing the general good above personal preference or comfort. This was necessary for the exercise of self-government, management of the family, and the stability and unity of the nation.

Work On the Self

In the discourses of these times, a woman had to work on herself to maintain her health and prevent illness. A woman could not do useful work or carry out her sacred duties if she were forever "sickly." Beecher's suggestions on how to care for the self included a balance among competing duties. Domestic manuals abounded, and advice on housekeeping, midwifery, childrearing, and sexuality replaced formal medical treatises. Longer, more discursive corporeal regimes with advice for diet, managing the emotions, specific illness related remedies, and "the sexual needs and anxiety of a growing middle class" began to replace physician-authored textbooks (Rosenberg, 1998). Advice manuals in the early and mid decades of nineteenth century America inscribed certain truths about the body in the same way that health textbooks do today. In this sense health advice manuals function as a mode of subjection "inscribing a particular relation to oneself" and to others through the inculcation of certain regimes of the body (Rose, 1996a, p.137). Women were responsible for "catch[ing] children when they were young" and instilling health habits as a condition for prevention. It is "better to save children from being poisoned, than to pay physicians for trying to cure them after they are contaminated, and, in many cases beyond the reach of a cure" (Beecher, 1845, p. 81). These manuals had a broad circulation in antebellum America and were an important pedagogical device for circulating health advice. These health discourses inscribed subjectivities or the "making up" of certain kinds of individuals that are different from today's health promotion discourse (Hacking, 2003).

Catharine Beecher wrote one such advice manual that was used in early common schools (Woody, 1929). This text was designed to cultivate particular "technologies of the self" (Foucault, 1988, p. 18).[6] These are a cluster of practices that individuals actively use to work upon themselves—to fashion themselves as particular kinds of beings. Beecher's text, *A treatise on domestic economy* (1841/1977), was concerned with inculcating rules to follow in order to manage a home that was efficient, orderly, and healthful for the family.[7] Instilling rules of conduct or habits for character formation and healthful habits was at the heart of this text.

Beecher imagined the home as a place with regenerative qualities and gentle Christian virtues, where love and kindness are instilled and nourished. Each chapter began with a set of "principles." For example, in the chapter "Management of Young Children," mothers who were in close contact with their children on a daily basis were advised to express sympathy, encouragement, and tenderness in their governance rituals. Mothers were cautioned to "*advise* and *request*, rather than command" [original italics] young children, and noted that "the little acts of heedlessness, or awkwardness, or ill-manners, so frequently occurring with children, should pass as instances of forgetfulness, and not as acts of direct disobedience" (p. 230). She noted that these "habits" were difficult to carry out at times, but with careful attention to her own actions and feelings, the mother would learn self-control.

Self-control was an important virtue to nurture, as the happiness of the family depended upon the "cheerful temper and tones". . . [of] the housekeeper. . ."which . . . renders it easier for all to do right, under her administration" (p. 134). The "government of the tones of voice" was a principle mothers needed to instill in themselves to ensure the "comfort and well-being of the family." A woman who went about "her house with a . . . . stinging snapper more than destroys all the comfort that otherwise would result from her system, neatness, and economy" (pp. 134–138). Mothers needed to "cultivate these habits" in themselves before they could expect to instill these habits in their children (pp. 140–141), for while it was

> desirable that children grow up in habits of system, neatness and order; . . . it is still more important, that they grow up with amiable tempers, that they learn to meet the crosses of life with patience and cheerfulness; and nothing has a greater influence to secure this, than a mother's example (p. 138).

Each chapter ended in a verse from a prayer or the Bible, such as: "forgive us our trespasses, as we forgive those who trespass against us," which Beecher noted that "every parent . . . needs daily to cultivate the spirit" expressed in [this] Divine prayer (p. 140). Women were to "save" the nation by honoring their "peculiar" nurturing duties as nurses, teachers, and mothers, roles thought to be congruent with the female design set by nature (Beecher, 1865). Morality was not separated from practices of the self. Instead, practices found their "truth," or justification, in the newly constructed morality of motherhood/wifehood as savior. Children learned through imitation—reiterating the truth of "self-evident" practices of the redeemed and redeeming mother: "It is in her hand that first stamps impressions on the immortal spirit." (Beecher, 1865, p. 6).

A Gendered Curriculum

And . . . it is true, that the education necessary to fit a woman to be a teacher, is exactly the one that best fits her for that domestic relation she is primarily designed to fill. (Beecher, 1835, p. 18)

Beecher's educational program in *An essay on the education of female teachers* (1835) imagined a gendered curriculum for seminaries and common schools that would forge "uniformly well educated pupils." As schools were to prepare students for their future roles, Beecher crafted a particular function for young girls defined in the language of duties. Beecher asks, "What is the most important and peculiar duty of the female sex?" and responds with "It is the physical, intellectual, and moral education of children. It is the care of the hearth, and the formation of the character of the future citizen of this great nation" (p. 5). Women were particularly fitted for the role of teachers because of their presumed sentiments, character, and disposition for the "government and education of the various characters and tempers that meet in the nursery and school-room." Beecher asserts that there had been a change in the education of females and that "mental discipline," which had been the privilege of men only, was beginning to exert its influence on the character of females. So, "Instead of the fainting, weeping, vapid, pretty play-thing, once the model of female loveliness, those qualities of the head and heart that best qualify a woman for her duties, are demanded and admired" (p. 6). The subject matter that was deficient in schooling was that of moral and religious training, as these were thought to have a "decided influence" in forming the character, and regulating the principles and conduct, of future life" (1835, p. 7).

In this new understanding of womanhood, women had a duty to self, God, family, others, and the nation.[8] An ailing mother created anxieties, as she was a constant threat to the unity and stability of the new republic. Beecher insisted that the primary means of exercising these duties was through the practice of self-denial. Through the virtue of self-denial came the possibility of constituting a new human nature, free of sin (disease, delinquency). Truth, previously of scripture, was now through science. The Word no longer passed through the body of Jesus as Holy Spirit; it is now passed as science through the body of the essence "mother/wife." Scripture was scarcely less prominent for it was constantly applied to the great variety of practical subjects discussed. Justification for rules relied on the Scripture. It was the Holy Spirit, reconstituted as the technology, "mother/wife" that articulated technologies of the self to the people on earth.[9]

Beecher's contemporaries also discussed household management in terms of a science in which women needed to receive proper training to complete tasks efficiently:

> Other things being equal, a woman of the highest mental endow-ments will always be the best housekeeper, for domestic economy, is a science that brings into action the qualities of the mind, as well as the graces of the heart. A quick perception, judgment, discrimination, decision and order are high attributes of mind, and are all in daily exercise in the well ordering of a family . . . The influence of women over the minds and character of children of both sexes, is allowed to be far greater than that of men. This being the case by the very order-ing of nature, women should be prepared by education for the per-formance of their sacred duties as mothers. (Grimké, 1837)

Whether by design or circumstance, it was reasoned that women were responsible for the well being of their families, society, and the race. Some contemporaries of Beecher thought of her pedagogical plan as repressive—designed to socialize women into conventional roles within the domestic space—as wives and mothers. Instead, Beecher's new scheme was produc-tive. It produced a new way of thinking about how girls were to be edu-cated, especially as it related to health issues. Female bodies were now constructed as useful, robust, productive, and vigorous, not passive, frail, and sickly. Beecher's claim over a particular kind of education was a marked one, one that did not play down embedded cultural sexual dualisms. On the one hand, gender boundaries were made more secure, yet on the other hand some borders were breached. Women and girls could now think differently about their bodies' capacities and capabilities: they could now conceptualize their bodies as vigorous, active, and capable, characteristics that during this timespace were considered "masculine"(Borish, 1987). The new pedagogy stressed traits that both men and women could aspire to.

Health reform discourse brought a structured regime and new way to think about life, along with the promise of a better life through its programs in education. As noted by Foucault, "power . . . was . . . taking charge of life, [it was life] more than the threat of death, that gave power its access even to the body" (1990, pp. 142–143). Life itself was made an object of knowledge, thus increasing the possibilities for intervention, regulation, governing, and ever fine-tuning of life processes. This power is a cultivat-ing form of power and calculates how best to produce life through gener-ating norms of existence. Fields of knowledge concerned with life and life processes, such as biology, anatomy, and especially physiology were being established as scientific disciplines and are examples of this new possibility to order and regulate life (Daniels, 1996). Knowledges from these academic disciplines were taken up as justifications for specific health advice in relation to age-specific schooling practices. For instance, Jackson (1845) argued, based on the "science" of physiology, "that an hour's confinement should be followed by a recess of fifteen minutes. In very young children (three to five), the period of confinement should be shorter, and the recess longer" (¶ 3–4). We can think of physiology as a way of thinking or a mode for self-improvement. Physiology, in all its forms provided a set of categories in which individuals could make sense of their own and others lives. With its classifications of the body systems and anatomical drawings, it functioned as a pedagogical devise for reformers.

Beecher's *Treatise on domestic economy* (1841/1977) is an example of the "hygienic regulation of domestic life" (Rose, 1996b, p. 6). Beecher (1865) constructed women as the family's "chief ministers." The well being of the family was her most important duty—to herself, to God and to the nation, and as such, her own health was of paramount importance. Whether talking about how women should think about the spatial layout of the home to ensure adequate sunlight and ventilation, clothing, food, gardens, or amusements, Beecher's central concern was on matters of health—how the body functioned and how to keep it healthy in order to fulfill her duties. The home, as was the body, was the temple of God. Beecher targeted mothers as the "moral" agent of the home—as the primary nurturer and educator of the nation's youth. Most of the writing concentrated on what was good/evil, efficient/wasteful use of energy, useful/not useful (frivolous) and what was allowed/not allowed. She set forth general principles and rules for conduct.

In discussions related to health, her reasoning concentrated on physiological explanations of how the body functioned or was degraded (degenerated) or made unhealthy by not following the "laws of health," which, for Beecher were the "laws of God" (1856, p. v). For example, in explaining the "first cause of mental disease and suffering," Beecher noted that the cause was for

> want of proper supply of duly oxygenized blood. It has been shown, that the blood, in passing through the lungs, is purified, by the oxygen of the air combining with the superabundant hydrogen and carbon of the venous blood, thus forming carbonic acid and water, which are expired into the atmosphere. (1842, p. 197)

At times, appeals to expert knowledge evidenced by extensive passages from "a distinguished medical gentleman" or from physicians (e.g., Dr. Combe) were included in her writings about health advice, such as in the chapter on infant feeding. According to Dussel (2001), a new culture of the body was emerging in the eighteenth and nineteenth centuries. Treatises that had previously concentrated on the cultivation of manners, civilities, and codes of elegance had shifted to textbooks on hygiene. The body was thought of as an organic machine, whose functions could be explained scientifically. The body had to work to improve its strength. Therefore, exercise was needed and came to be valued in the "arts of training the body."

For Beecher, instilling good manners was also important. Manners were linked to good health: "Good manners are the expressions of benevolence in personal intercourse . . . the exterior exhibition of the Divine precept." (1842, p. 137). The "defects" of deportment in national manners were to be corrected in the home and in the schools to prevent the degeneration of the individual and the nation. It was these manners and deportments of the "wealthy" that Beecher deplored, and which she asserted others—"those of lesser means" were aspiring to, as they were unhealthy. In attempts to be "fashionable," all sorts of unhealthy "aristocratic" behaviors were undertaken by young girls and women. Practices such as dressing inappropriately for the climate, wearing tight shoes not meant for walking, sitting during long piano sessions and serving of dainty foods other than at mealtime were, among others, habits that needed to change.

In *Suggestions respecting improvement in education* (1829), Beecher noted that "The improvements made have previously related chiefly to *intellectual acquisitions*," [original italics] but this is not the most important object of education.

> "[T]he correction of the disposition, the regulation of social feelings, the formation of the conscience, and the direction of the moral character and habits, are united, objects of much greater consequence than the mere communication of knowledge and the discipline of the intellectual powers. (Cited in Sklar, 1973, p. 91)

Gaining "knowledge of the construction of the body and the laws of health" was seen by Beecher as a "matter of duty" (1856, p. v). Beecher sought to make the "laws of health" into self-governing practices through their dissemination in school textbooks.

All Things in Balance

Beecher privileged the principle of balance as a mode of ensuring health. This came into play in many forms, including the necessity to balance physical work with intellectual work. There was to be balance between "stimulation of the intellect" and what Beecher called "the physical and domestic education." As a rule, girls should not start school before the age of six, and should not be confined to any employment for more than one hour, which was then to be followed by "sports in the fresh air." Young females needed protection from too much "mental stimulation" in their early years of development. This may seem a contradiction when many, including Beecher, saw the lack of schooling for girls as a major problem. But her reasoning was that the period or "stage" between the ages of six and 14 or 15 was "the most critical period of [a girl's] youth." During this time young girls were establishing a "vigorous and healthful constitution" and if too much "intellectual excitement" was demanded of young women, it was presumed to lead to decay.

Putting too much stress on "intellectual culture" and little, if any, on "physical development" was a particular problem of those belonging to the wealthier class. It was the mothers' "duty" to "set a proper example" and to "make it their first aim to secure a strong and healthful constitution for their daughters, by active domestic employments" (1842, p. 50). Her first suggestion for females was to have only afternoon classes so that the mornings "might be occupied in domestic exercise" such as sweeping, dusting, care of furniture and beds, washing and cooking, "which should be done by the daughters of a family."

If unable to guarantee daily "domestic exercise," Beecher recommended developing a daily routine in calisthenic exercises. For Beecher, calisthenics was a remedy for various ailments, a device useful "as a mode of curing distortion, particularly all tendencies to curvature of the spine;" while, at the same time, it tends to promote grace of movements, and easy manners." These exercises "combined with music" were to secure "all the advantages which dancing is supposed to effect, and, which is free from the dangerous tendencies of that fascinating and fashionable amusement" (1842, p. 56).

Planned exercises, either "domestic" or "physical" had to be useful in strengthening the body. Exercises also had to be efficient in nature, so as not to deplete body reserves, which was in line with medical thinking. In the early and mid decades of nineteenth century America, both medical and lay thought conceptualized disease as disorder, an imbalance of fluids within the closed system of the body; it was a changed state of being that needed to be brought back into balance through various interventions. Exercising in excess, whether physical or mental, led to an imbalance in the body systems that could cause nervousness or other physical ailments. Moderation, the key to preventing imbalances, was the governing principle for how to live a morally responsible life. For Beecher, women's duties were to be carried out in moderation; too much time devoted to benevolent duties at the expense of duties to the family or self was wrong.

Beecher introduced calisthenics (female gymnastics) as a means of daily exercise geared toward the corporeal regulation of children's bodies. These exercises were brought into curricula to improve the health status of girls and to instill good habits. In her textbook *Physiology and calisthenics for schools and families* (1856), Beecher understood the female sex to be the primary site for nation building. Beecher depicted women's sexual/reproductive organs as a site of racial degeneration and regeneration. The "monstrous deformity" of women (produced by corset-wearing), Beecher claims, is "perpetuated through a degenerate offspring" (p. 21). She again reiterated her claims of the decreasing health of females and advocated calisthenics as a remedy; "American women every year become more and more nervous, sickly, and miserable, while they are bringing into existence a feeble, delicate, or deformed offspring" (p. 151). Reasoning about women's "usefulness" primarily through their reproductive capacity, traveled across disciplines and were common discourses in the popular health movement. Beecher's calisthenic exercises resembled meticulous, precision military drills designed to discipline the corporeal body in the name of health.

The regime of domestic or physical exercise was accompanied by the introduction of the study of basic physiology. Beecher's textbook functioned as a physiology primer with detailed drawings of the different body systems, such as muscles, nerves, bones, and organs. These illustrations were part of the didactic learning that she advocated for women and girls in order to understand the structure and functioning of the body. The upsurge in the use of texts such as these also had other effects. With the increased importance of scientific knowledges such as physiology by health reformers, bodies became marked in an increasingly discursive construction of differential male and female bodies. As noted previously, health discourses construct both the subject matter and the individuals subjectivity. Physiological thinking led to the logic that since males and females were so irrefutably different, so must their roles be different. These were simultaneously configured into traditional roles reflecting the private and public spheres of activity of the sexes. However, the question arises: How are these roles themselves constructed? Women's roles themselves were largely created through this very discourse, and in this timespace they were justified through the spiritualization of the science of physiology.

> The male organism, including body and soul, is adapted to elaborate, secrete and impart the primary element, or germ, of a new being; the female is adapted to receive, nourish and develop that germ into a living human form. In the masculine organism, the seed is formed; in the feminine, it must be nourished" . . . This distinction has a "universality" [which] marks all species, plants, animals and man. (Wright, 1854, pp. 20–21)

The heritability of character traits as well as physiological traits was a constant theme in health discourses: "Children not only inherit goods and houses and money from their parents, "but also their *bodies and their souls*" [italics in the original] (Wright, p. 16). The womb functioned in a nutritive role in providing nourishment to the babe. This nutritive function does not stop at birth of the baby but extends into its life. By extension, the mother would continue to provide a nutritive role both in the home and in society.

By the 1860s, new ways of thinking about health and the body were emerging. There was a shift from thinking about the body order and health: from that of ideas of the "natural" to those of the "normal." Rather than looking at health as a natural balance among the body, mind, and soul, physicians began to look at health statistically in terms of conformity or deviations from a physiological norm. Warner (1990) notes that physicians had come to think about bodily problems less as systemic imbalances in the body's natural harmony and more as complexes of discrete signs and systems that could be analyzed, separated, classified, and measured in isolation. Warner argues that the vocabulary of health shifted dramatically in a few short years. These permutations were evident in hospital case histories. Specific indicators of health, such as pulse or the chemical composition of urine, were weighed against criteria of health and formulated as norms for a population or as universalized norms defined by laboratory science. Graphs of quantified indicators of health became commonplace in medical records. The body came under the control of a mathematization of its functions. Subjectivities could be articulated, categorized, and measured for normalization under scientific study. These new procedures, as described by Warner, opened new possibilities for the governing of the self.

Conclusions

The first half of the nineteenth century can be viewed as a distinct discursive timespace. Health had emerged as a problem and people now realized that they could take preventive measures to improve their health and to prevent debility. This was primarily taken up through the work on the self. The newly emerging sciences such as physiology became important, in that this knowledge provided authoritative rationales for health reformers. Health content, in the form of physiology and anatomy—how the body functioned—was privileged in the antebellum curriculum as it is in today's health curriculum.

Beecher's texts are important today in that they laid out principles for living a healthful life as do the health standards we have today. Both timespaces involve the double logic related to the making of legible, manageable health practices and the fabrication of individuals with certain dispositions. During this timespace, mothers emerged as the primary persons to educate their children in the home and the notion that women were best suited to educate the child took hold in common schools. Today's health textbooks function as translating devices between the experts of the medical world and governmental health agencies and teachers and children. These textbooks take the medical jargon of the disparate world of medicine and science and make it intelligible to schoolchildren in a developmentally age-appropriate manner. While today's health texts retain some of the historical continuity or the "living on" of authoritative tomes/messages, they do not project the authoritarian and moralistic tone of health advice of nineteenth century America.

Since the 1950s, other shifts in how we think of health have emerged. Currently, the idea of a protoprofessional is used by de Swaan (1990) in explaining how individuals have become their own experts, using the vocabulary of science and readily entering into therapeutic relationships with all sorts of professionals to address everyday lifestyle decisions. We no longer need strict "disciplinary" regulations imposed upon us. As noted by Rose (1999), we now freely enter these relationships. Today's health promotion discourses position the individual as a self-monitoring, self-reflexive desiring subject. The subjectivities constructed in health education discourses in the decades of mid-nineteenth century and the subjectivities constructed in today's "health promotion" are both considered in their own time to be normal. I have sought to illustrate that "normal" is not a constant condition of nature. Rather, it varies with the cultural construction of subjectivities in relation to the historical conditions, needs, and desires of the times. In each case, the constructions delineate inclusions and exclusions. One effect is that in describing ourselves in terms of norms, it establishes classifications for groups who do not "fit" these norms. These groups become the excluded or the "other." In Beecher's timespace, the not normal were sin, disease, death, and decay. In the present, normal is constituted as informed choice through access to expert information and the shaping of healthy desire. Thinking in terms of the normal and "risk factors" has become part of our vocabulary. We have come to think of managing health risks, not the actual illness (Castel, 1991) in the care of the self.

Notes

★ I use the term "timespace" in this chapter to convey that what it means to be healthy is local and contingent, specific to diverse times and places.

1. For example, catechisms can be viewed as an early form of standard (see Rosenberg, 1995).
2. Foucault's use of governmentality or government is to disassociate it from conceptions of the state. Instead, it is concerned with a range of everyday practices, tactics, techniques, desires, and programs that help shape and regulate the self and others.
3. Whether women were actually more sickly than their Old World counterparts is not my major focus, but most reformers agreed that the health of the individual as well as the nation's health was progressively becoming more degenerate: "the human frame was degenerate, [and] had declined from its former glory" (Whorton, 1982, p. 18).
4. Tolley argues that education in the sciences (botany, chemistry) were implemented into the curriculum in female schools in the early decades of the nineteenth century while schools for males primarily taught the classics (1996).
5. Rosenberg (1995) notes that the 1830s were landmark years in that textbooks began to be published for use in common schools and the home relating to anatomy, physiology, and hygiene.
6. According to Foucault, such technologies permit individuals to effect, by their own means, or with the help of others a certain number of operations on their own bodies, and souls, thoughts, conduct, and way of being, so as to transform themselves in order to attain a certain state of happiness, purity, wisdom, perfection or immortality (1988, p. 18).
7. This text went through many editions and reprints. Sklar noted that it served as a "scientific but personal guide" related to "health, diet, hygiene, and general well-being" (1973, p. 154).
8. Mann (1853), active in school reform argued that women had "divinely-adapted energies that were useful in the work of regenerating the world" (p. 67).
9. For example, Acts 15; 28–29, Romans 8: 1–11, I Corinthians 12: 1–11, Galatians 5: 16–26. (see *Good News Bible: Today's English Version*, 1976, New York: American Bible Society).

References

Armstrong, D. (1998). Bodies of knowledge: Knowledge of bodies. In C. Jones & R. Porter (Eds.), *Reassessing Foucault: Power, medicine and the body* (pp. 17–27). London: Routledge. Bartlett, E. (1838). *Obedience to the laws of health: A moral duty.* Boston, MA: Julius A. Noble.

Beecher, C. E. (1829). *Suggestions respecting improvements in education.* Hartford, CT (Excerpts pp. 7–16). Retrieved July 23, 2003, from http://www.women.eb.com/pri/Q0017.html

Beecher, C. E. (1835). *An essay on the education of female teachers.* New York: Wm Van Norden.

Beecher, C. E. (1842). *A treatise on domestic economy, for the use of young ladies at home, and at school* (rev. ed.). Boston, MA: T. H. Webb.

Beecher, C. E. (1845). *The duty of American women to their country.* New York: Harper & Brothers.

Beecher, C. E. (1856). *Physiology and calisthenics for schools and families.* New York: Harper Brothers.

Beecher, C. E. (1865). Ministry of women. In B. M. Cross (Ed.), *The educated woman in America: Selected writings of Catharine Beecher, Margaret Fuller, and M. Carey Thomas* (pp. 94–101). New York: Teachers College Press.

Beecher, C. E. (1977). *A treatise on domestic economy.* New York: Schocken Books. (Original work published 1841)

Blackwell, E. (1852). The laws of life, with special reference to the physical education of girls. New York: George P. Putnam.

Bloch, R. E. (1992). American feminine ideals in transition: The rise of the moral mother, 1785–1815. In N. F. Cott (Ed.) *History of women in the United States: Historical articles on women's lives and activities, domestic ideology and domestic work* (Vol. 4, Part 1) (pp. 3–28). New York: K. G. Saur.

Borish, L. J. (1987). The robust woman and the muscular Christian: Catharine Beecher, Thomas Higginson, and their vision of American society, health and physical activities. *The International Journal of the History of Sport 14,* 139–154.

Breslow, L. (1999). From disease prevention to health promotion. *Journal of the American Medical Association, 281*(11), 1030–1033.

Castel, R. (1991). From dangerous to risk. In G. Burchell, C. Gordon, & P. Miller (Eds.) *The Foucault effect: Studies in governmentality* (pp. 281–298). Chicago, IL: University of Chicago Press.

Channing, K. (1994). Feminist history after the linguistic turn: Historicizing discourse and experience. *Signs 19* (2), 368–404.

Daniels, G. H. (1996). The process of professionalization in American science: The emergent period, 1820–1860. In R. L. Numbers & C. E. Rosenberg (Eds.), *The scientific enterprise in America: Readings from Isis* (pp. 21–36). Chicago, IL: University of Chicago Press.

Dean, M. (1994). *Critical and effective histories: Foucault's methods and historical sociology.* New York: Routledge.

De Swaan, A. (1990). *The management of normality: Critical essays in health and welfare.* Routledge: New York.

Dussel, I. (2001). School uniforms and the disciplining of appearances. In T. S. Popkewitz, B. M. Franklin, & M. A. Pereyra (Eds.) *Cultural history and education: Critical essays on knowledge and schooling* (pp. 205–241). New York: RoutledgeFalmer.

Foucault, M. (1983). The subject and power. In H. L. Dreyfus & P. Rabinow, *Michel Foucault: Beyond structuralism and hermeneutics* (pp. 208–229). Chicago, IL: University of Chicago Press.

Foucault, M. (1988). Technologies of the self. In L. Martin, H. Gutman, & P. Hutton (Eds.) *Technologies of the self* (pp. 16–49). London: Tavistock.

Foucault, M. (1990). *History of sexuality: An introduction,* Vol. 1 (R. Hurley, Trans.). New York: Vintage.

Gastaldo, D. (1997). Is health education good for you? Rethinking health education through the concept of bio-power (pp. 113–133). In A. Petersen & R. Bunton (Eds.), *Foucault, health and medicine.* New York: Routledge.

Grimké, S. (1837). Letters on the equality of the sexes. *Letter VII, Condition in some parts of Europe and America.* Retrieved October 22, 2004 from http://www.pinn.net/~sunshine/book_sum/grimke3.html

Hacking, I. (2003). *Historical ontology.* Cambridge, MA: Harvard University Press.

Jackson, J. (1845). Confinement of children in school (from *The Mother's Assistant,* July, pp. 7–8). Retrieved December 11, 2004, from http://www.merrycoz.org/articles/CONFINE.HTM

Joint Committee on National Health Education Standards (1995). *Achieving health literacy: An investment in the future.* Atlanta, GA: American Cancer Society.

Lupton, D. (1995). *The imperative to health.* London: Sage.

Mann, H. (1853). *A few thoughts on the powers and duties of woman: Two lectures*. Syracuse, NY: Hall, Mills, and Company.

Meeks, L., Heit, P., & Page, R. (2003). *Comprehensive school health education: Totally awesome strategies for teaching health* (3rd ed.). New York: McGraw-Hill.

Osborne, T. (1992). Medicine and epistemology: Michel Foucault and the liberality of clinical reason. *History of Human Sciences 5*(2), 6393.

Our daughters (1857, Dec.). *Harper's New Monthly Magazine*, Vol. 16, no. 91, (pp. 72–78).

Petersen, A. R. (1996). Risk and the regulated self: The discourse of health promotion as politics of uncertainty. *ANZ Journal of Surgery 32* (1), 44–57.

Popkewitz, T. S. (2004). Educational standards: Mapping who we are and are to become. *The Journal of the Learning Sciences 13* (2), 243–256.

Rose, N. (1996a). Identity, genealogy, history. In S. Hall & P. duGay, *Questions of cultural identity* (pp. 128–150). Thousand Oaks, CA: Sage.

Rose, N. (1996b). *Inventing our selves: Psychology, power, and personhood*. Cambridge: Cambridge University Press.

Rose, N. (1999). *Powers of freedom: Reframing political thought*. Cambridge: Cambridge University Press.

Rosenberg, C. E. (1995). Catechisms of health: The body in the prebellum classroom. *Bulletin of the History of Medicine 69*, 175–197.

Rosenberg, C. E. (1997). *No other gods: On science and American social thought* (rev. and exp. ed.). Baltimore, MD: The Johns Hopkins University Press.

Rosenberg, C. E. (1998). *The book in the sickroom: A tradition of print and practice*. Retrieved October 15, 2004, from http://www.librarycompany.org/doctor/rosen.html

Sklar, K. Kish (1973). *Catharine Beecher: A study in American domesticity*. New Haven, CT: Yale University Press.

Tolley, K. (1996). Science for ladies, classics for gentlemen: A comparative analysis of scientific subjects in the curricula of boys' and girls' secondary schools in the United States, 1794–1850. *History of Education Quarterly 36* (2), 129–153.

U.S. Department of Education (2002). *The No Child Left Behind Act of 2001: Executive Summary*, Washington, DC. Retrieved May 7, 2003, from http://www.ed.gov/nclb

Warner, J. H. (1990). From specificity to universalism in medical therapeutics: Transformation in the nineteenth-century United States. In Y. Kawakita, S. Sakai, & Y. Otsuka (Eds.), *History of Therapy: Proceedings of the 10th International Symposium on the comparative history of medicine—East and West* (pp. 193–224). Tokyo: Ishiyaku EuroAmerica.

Whorton, J. C. (1982). *Crusaders for fitness: The history of American health reformers*. Princeton, NJ: Princeton University Press.

Woody, T. (1929/1966). *The history of women's education in the United States* (Vol. 1). New York: Octagon Books.

Wright, H. C. (1854). *Marriage and parentage: The reproductive element in man, as a means to his elevation and happiness*. Boston, MA: Bela Marsh.

CHAPTER 6

How Might Teachers of Young Children Interrogate Images as Visual Culture?

NANCY PAULY

Visual images, and experiences of seeing and being seen, saturate the public and private spaces where children learn to construct sociocultural and historical meanings about their identities, histories, and cultural values. Images permeate children's culture, appearing in TV shows, music videos, interactive games, fast food promotions, movies, videos, books, and various forms of visual art such as painting, sculpture, and architecture. While visual images have emerged in the last century as one of the most pervasive forms of human communication and persuasion, their enormous power as cultural texts used by children is largely ignored in educational discourse.

This chapter addresses the need for preservice teachers to interrogate images as powerful modes of communication, which participate within discourses of meaning and power that have real consequences in children's lives. Images, in mass media, popular culture, the school environment, and home cultures, are commonly produced to influence children to become consumers of specific ideas, objects, and experiences. This chapter explores how teacher education students gained access to memories from their early childhoods by deconstructing images as sociocultural texts and constructing websites for teaching. These experiences gave them the possibility of thinking in new ways, as well as imagining approaches they could use in their teaching with young children to create alternate ways of questioning, understanding, and perceiving reality.

This research offers new tools derived from the literature of Visual Culture Studies, which draws from postmodern, poststructuralist, and postcolonial theorists in disciplines such as cultural studies, philosophy, art history, art education, and media studies. First, there is a discussion of visual images as sociocultural texts. Second, the author discusses the ways that children negotiate meanings based on culturally-learned codes of representation and the ways that visual images may be linked to cultural narratives.

Seven approaches to culturally-based image interpretation are recommended to help preserve teachers investigate meanings. These approaches include culturally-learned codes of representation, interpreting images intersubjectively, sociocultural-historical contextualization, intertextual articulation, cultural narratives/discourses of knowledge and power, potential social consequences, and response-ability.

The writings by two preservice teachers, which are based on these approaches, follows. One of the participants is a Japanese-American woman who reflected on the image of a Japanese man in *Breakfast at Tiffany's* based on her own experiences and the history of Japanese internment. The second is a European-American man who explores *G.I. JOE®* and asks, "How do guns shape male identification in our culture?" This is followed by a website that he constructed to help children examine images of violence.

Finally, this author recommends that teachers and young students investigate images in multiple ways through discussion, art making, writing, and drama to discover ways that images, when linked with cultural narratives, invite children to think, feel, act, and imagine themselves and others in sociocultural terms.

Visual Images as Sociocultural Texts

Since the earliest cave paintings 21,000 years ago in southern Africa, humans have created visual images throughout the world to encode their values and experiences in metaphoric forms. Often viewers of visual art and artifacts are engulfed in meaningful sociocultural experiences that help them to understand themselves as members of societies. For example, during the Middle Ages, images on paintings and stained glass windows functioned to educate and inspire people in their faith, while large buildings or metal sculptures commemorated the power of their rulers or wealthy merchants. Similarly, in West Africa the Yoruba people have historically created and performed masks with elaborate costumes, prayers, songs, and drums that contribute to sociocultural meanings within their communities for hundreds of years. Individual artists have also expressed their particular feelings, experiences, and ideas within their cultural contexts.

Images as social texts changed radically during World War I when professionals in advertising, news, and illustration were employed by the government to sell the war. Freudian and Pavlovian psychologies offered them new tools to motivate behavior unconsciously through a learned stimulus-response process. Public campaigns commonly offered emotional and symbolic images and stories, rather than facts, to convince citizens to join the war efforts. The film industry, and later television, offered new forms of mass communication. Advertisers learned they could symbolically link images and objects with our most basic desires for love, security, patriotism, or pleasure through the repetition of images linked with cultural narratives. After enough conditioning, an image could trigger positive or negative associations without supporting texts that would motivate people to act; to buy a hamburger or go to war.

Now commercial images are commonly produced to sell children and adults everything from toys and fast food to politicians, histories, and identities by linking visual images and objects symbolically with cultural narratives and basic human desires. Many advertisements construct citizens as consumers. The role of the citizen/consumer was well illustrated soon after the terrorist attacks on September 11, 2001, when President Bush immediately advised citizens to buy products to show their patriotism.

Commercial arts usually hide the contexts, sources, affects, and possible consequences of consumerism. Images suggest that consumerism is something beautiful, fun, and desirable. For example, McDonald's® current advertisements for children usually show "popular" kids skateboarding and laughing together while the theme song and texts repeat, "i'm lovin' it." McDonald's is selling the experience of friends, popularity, and fun through their food. Children are one of the primary targets of advertising today. Advertisers talk about "branding" children as young as possible to be lifelong customers.

It is very important for teachers and children to consciously learn how to investigate images for artistic metaphors that may enrich their lives as well as messages that may have a variety of social consequences. Viewers can consider one image to investigate: the culturally learned codes of representation, their own subjectivities, their associations with an image, the historical contexts within which an image emerged, the sources and motivations that made that image possible, the potential effects that image may suggest, and ways people may respond to the messages that images imply through their actions.

Many psychologists, such as Wertsch (1991) and Rogoff (2003), have used the writings of Vygotsky and his colleagues to understand individual development within social, cultural and historical contexts. "Through engaging with others in complex thinking that makes use of cultural tools of thought, children become able to carry out such thinking independently, transforming the cultural tools for thought to their own purposes" (Rogoff, 2003, p. 50). Our culture has changed dramatically and now uses images to educate and persuade its young through television, video, the internet, interactive video games, and the like. Educators must adapt by providing children with new cultural tools to interpret visual images using approaches based on sociocultural theories.

Children Negotiate Meanings Based on Culturally
Learned Codes of Representation

In the social constructivist perspective based on Hall (1997, p. 61), representation involves making meaning by forging links between three different orders of things: (1) what we might broadly call the world of things, people, events, and experiences; (2) the conceptual world—the mental concepts we carry around in our heads; and (3) the signs, like images and words, that are arranged in languages, which stand for, or communicate, these concepts.

Representation is both a concept and a set of practices by which meanings are produced and exchanged between members of cultures. Words and images stand for things that producers encode and viewers decode based on the conceptual maps that we carry around in our heads (Hall, 1997, p. 15). People from the same cultures usually interpret meanings similarly, based on their shared histories, values, and codes of representation, yet their interpretations may vary based on the contexts within which meanings are learned and the needs of individual users at particular times.

Visual images are commonly used by children, in conjunction with other signs and symbols, to understand their cultures and "make sense" of their positionalities and possibilities within cultures.

> Culture . . . is not so much a set of things . . . as a process, a set of practices. Primarily, culture is concerned with the production and the exchange of meanings—the "giving and taking of meaning"—between the members of a society or group . . . Thus culture depends on its participants interpreting meaningfully what is happening around them, and "making sense" of the world, in broadly similar ways. (Hall, 1997, p. 2)

Visual Images as Linked to Cultural Narratives

This research suggests that children and adults often learn to link visual images to cultural narratives, or social stories, through our memories or associations, which they unconsciously use to interpret meanings. The term "cultural narratives" is derived from Friedman (1998) who writes,

> [c]ultural narratives encode and encrypt in story form the norms, values, and ideologies of the social order . . . around which institutions of gender, race, class, and sexuality are organized. Cultural narratives also tell the strategic plots of interaction and resistance as groups and individuals negotiate with and against hegemonic scripts and histories. (Friedman, 1998, p. 8)

Cultural narratives may involve stories about racism, ethnicity, sexual identity, class, gender, body image, beauty, psychological dispositions, social acceptability, citizenship, consumerism, militarism, nationalism, imperialism, traditionalism, modernity, technological superiority/inferiority, or other aspects of human experiences.

Images associated with cultural stories help students to imagine how they might imagine and perform their socially learned identities as individuals or citizens. Like the elements of a play or novel, images help children to imagine which characters they might play in life, which parts might be available to them, who has power and why, and what cultural narratives dominate this drama. Images participate within discourses that suggest the possible social consequences of challenging the script or playing their social parts in unexpected ways. They surmise how power works in relation to their bodies, and their social identities, by studying the images and stories connected with themselves or other people in mass media, popular culture, and their communities. In other words, they unconsciously learn who and what are valued in each telling of the cultural tale.

I am using the term Cultural Narratives somewhat interchangeably with the concept of discourse as discussed by Foucault (1980), Fiske (1996) and Gee (1999). Discourse implies ways that language and other cultural texts are used to construct understandings about the world that open particular ways of thinking and acting, while disinclining other possibilities.

Fiske (1996) described ways that language, media images, and practices are put into social use under *particular* historical, social, and political conditions (such as the O. J. Simpson trial) to frame discourses of meaning and power (in this case, racism in America) within discursive communities.

Gee (1996) personalized discourses when he declared,

> Discourses . . . are ways of behaving, interacting, valuing, thinking, believing, speaking, and often reading and writing, that are accepted as instantiations of particular roles (or "types of people") by *specific groups of people* . . . Discourses are ways of being "people like us." They are "ways of being in the world". . . and products of social histories (p. viii).

Both Fiske and Gee have been interested in the ways cultural texts circulate cultural meanings through discourse. While the concept of discourse as exemplified by Fiske, Gee, and Foucault ground my theory, I have preferred to relate images to the term "cultural narrative" rather than "discourses" because I believe that visual images are experienced as performed meanings, with the imagination, and through the body, and then are interpreted as parts of culturally rooted "movies or stories" in the mind that are more akin to dramatic enactments of narrative forms, rather than logical or rhetorical uses of speech.

I have proposed that children learn to associate images with cultural narratives as "image-narratives" (Pauly, 2002), which they employ as mental tools to unconsciously interpret new experiences. The same tools might be used to consciously unlock a variety of culturally learned meanings and consequences.

Children *learn* to link images with the stories. For example, Disney® films tell stories about male and female relationships, physical beauty, and ethnicity, among others. In terms of relationships, Snow White hides from an evil woman until a Prince rescues her. The Little Mermaid works with an evil woman to trade her identity as a mermaid, and her beautiful voice, for human legs so she can marry the prince, with whom she has never spoken. It is also interesting to note that no major Disney female character had a wise mother from whom she could learn.

Children often *perform* the messages they learn as Butler (1990) suggests. For example, a friend of mine (who is also a feminist art historian) told me that although she did not allow her daughter to watch Disney® films or buy Disney merchandise, her daughter had received "Little Mermaid" shoes from her cousin when she was four years old. One day my friend observed her daughter passively sitting in her bedroom wearing her new shoes and asked her what she was doing. Her daughter replied, "I am waiting for a man to come and save me." In this case, since the shoe fit, her daughter tried on the story. Needless to say, my friend discussed this image-narrative with her daughter.

Historically, the female Disney characters' bodies have changed over time from the body type of a young girl, like Snow White, to the buxom Little Mermaid and Jasmine who are girls with cleavage, an inhumanly small waist, and a scant midriff costume dipping well into the pelvis whose proportions are similar to Barbie® dolls.

Children can learn to find the messages and potential consequences within image-narratives and choose to challenge or *transform* the dominant messages found in art and popular culture. They can "talk back" (hooks, 1994) to images, make art to critique images, create art that is more relevant in their lives, or act in alternative ways. For example, Lisa, one of the preservice teachers in my study, interviewed Rochelle, a fifth grader, about Jasmine, a character from *Aladdin*®. Rochelle made a large drawing of Jasmine and wrote,

> What do I think? I [think] Jasman [Jasmine] is pathetic. Whats [*sic*] with the tiny little wast [waist] and amazing bewty [beauty]? I mean nobody looks like that. When I was little I used to try to look like that. Now, I know better. Most Disney women cariters [characters] have no personal[it]y what so ever. (Rochelle, personal interview by Lisa, preservice teacher, 1998)

Ethnicity is commonly constructed in exaggerated images and stereotypes. For example, the film *Aladdin*® contained racist descriptions in words and images such as the mispronunciation of Arab names, the stereotypic depiction of each character, and the nonsensical scrawl instead of Arabic writing. The introductory song begins, "Oh I come from a land/ From a faraway place/ Where the caravan camels roam. Where they cut off your ear/ If they don't like your face. It's barbaric, but hey, it's home." After protests from Arab Americans, such as Jack Shaheen, Disney® omitted the phrase "Where they cut off your ear/ If they don't like your face" in the videocassette but retained, "It's barbaric, but hey, it's home," as well as the other stereotypes.

I had a similar experience with the performance of an image-narrative with my nephew, Nicholas, when he was three years old. Nick and I were playing with a scene depicting the Nativity of Jesus. When I picked up the king with the dark complexion, Nicholas said, "He can't play with the other kings." When I asked him why, he responded, "These are daddies and that one is not a daddy." For three days, when we returned to our play, Nicholas ran away with the light-skinned kings when I moved the dark one toward them. On the third day, Nick and I watched *Aladdin®*, a video that the family had recently purchased. The story begins with an image of Jafar, the evil villain on horseback, and the narrator says, "This story begins with a dark man on a dark night with a dark purpose." Soon after we watched the video, I took the image of the dark-skinned king and asked Nicholas, "Is this Jafar?" Nicholas nodded positively. I explained that the king was not Jafar but the good king who brought gifts to the baby Jesus. Nick nodded. After that conversation Nicholas played with all the kings. When Nicholas was six, I asked him, "Do you remembered when we played with the kings at Christmas time?" He said, "yes." I continued, "Did you think the dark skinned king looked like Jafar?" "No," he responded, "I though he *was* Jafar." This story suggests that young children associate the general visual characteristics of a character with stories, and then reapply the image-narrative to people with similar visual characteristics. I asked Nick, who is now 13 years old, if he remembered playing with the nativity set when he was three. He said, sure, "I thought the king looked like Jafar, so I thought he was bad, and I wouldn't let him play with the other two kings" (Nick Pauly, personal correspondence, October 30, 2005).

Media theorists such as Kilbourne (1999), Ewen (1988), and Jhally (1987) have contended that people commonly interpret visual images unconsciously, as emotional or aesthetic experiences. Images engage our aesthetic pleasure or revulsion and evoke our fantasies, fears, or desires.

Although children seem to consume images passively, they actually *negotiate or construct* meanings based on their prior experiences, the context when the image is viewed, and their needs and desires at the time. They commonly link images with cultural stories that they may associate with "beauty," "truth," "goodness," or "normalcy" rather than contemplating the producer's intentions and the potential consequences of valuing the messages and images conveyed.

Since children commonly consume visual images unconsciously, this chapter recommends that preservice teachers explore the ways in which images are connected to their own ideas, experiences, feelings, histories, and desires to offer teachers ways to distinguish harmful messages from pleasurable sensations. Since teachers and students may interpret visual images differently, depending on their positionalities or experiences, it is important for adults or young children to explore the varied meanings that images may imply through discussions with a variety of people. These approaches are intended to provide tools that teachers might adapt in their future work with children.

Seven Culturally Learned Approaches to Interpreting
Visual Images

These seven approaches are neither discrete nor sequential methods but are rather analytic tools to interpret visual images as cultural texts. They were first discussed as six approaches in Pauly (2003) and are further elaborated here. From 1998 to 2003 I used questions that correspond to each approach as assignments in art education method courses with preservice elementary teachers and art teachers. Each of the preservice teachers was asked to select one visual image from art history, popular culture, mass media, children's books, a museum setting, or a living artist. I have included a description of the approaches and example questions to which the students responded and researched. Then I provide examples by two preservice elementary teachers.

Approach 1—Culturally Learned Codes of Representation

This approach acknowledges that representation occurs within cultural codes that are learned within cultures that have a history and a position within discursive formations, as outlined by Hall (1997). These codes of representation can be explored by looking at a visual image and simply asking the viewer, "What do you see?" If the viewer is viewing the image alone, he or she should list as many features as possible, then interview several other people (especially people whose social identities are different than their own) to hear other points of view. Ideally teachers and students should discuss the image in a group. The interpreter should search for details about the subject matter, signs, symbols, design qualities (such as color, repetition, and balance), technical qualities (such as the way the paint is mixed and applied), and other visual features that strike them. This approach is designed to notice the artwork carefully.

Artistic interpretation should also be a creative process in which the viewers *wonder* about the visual qualities of the work and open space for unexpected observations, reflections, and insights by noticing what is simply there to see, feel, or imagine. The visual arts, even the commercial arts, suggest meanings on many levels and cannot be reduced to single functions.

Next, the participant is asked if they can recall when they learned the meanings of these codes of representation. This approach is based on the work of Lanier (1982) and Hicks (1993) in art education who stress that people learn these codes through interaction with the people and media around them. Similarly, Tabachnick (1997) in teacher education has written:

> People learn in a social context: that is, they learn from their experiences with other people. We learn through the reactions of people to our ideas and our actions. We learn by thinking about people responding to our ideas even when those people are not physically present (p.1).

Culturally-learned codes of representation do not necessarily determine how a viewer interprets a visual image, but these codes do mediate and shape the availability of readings that an individual might use, as well as the political space they might need to voice alternatives. "The reading of any text cannot be understood independently of the historical and social experiences which construct how audiences interpret other texts" (Giroux, 1994, p. 200).

Approach 2—Interpreting Images Intersubjectively

Children learn the meanings of their culture *intersubjectively* (Moore, 1994, p. 3), that is, through interaction with other people and "cultural texts" (such as books, films, television, video games, the internet, restaurants, and stores) while they are recognized as having embodied social identities. Moore (1994), a feminist cultural anthropologist, maintains that identities are attached to our bodies learned through a "lived anatomy" (p. 3). We learn how we are socially identified in concrete spaces and times with other people or social texts.

Identity formation is based on belonging, or not belonging, to a group. People learn the meanings of their subjectivities, such as their genders, ethnicities, sexuality, religions, age, and class, while they are recognized, or not recognized, as belonging to that group. They learn the narratives and the expectations about "people like them," or those different than themselves, as embodied subjects.

When people view visual images they will unconsciously or consciously notice whether or not the image might refer to their bodies or to people they know. They will consider whether they are insiders or outsiders to the people, cultures, animals, or environments represented in images. Social identities and interpretations are not fixed but are constantly shifting based on each person's knowledge of history, personality, power relations, desires, and intentions at particular times.

Identities are encoded in words that have histories and meanings. For example, Stuart Hall (1989), a cultural studies theorist, was a Jamaican who moved to England in the 1950s. In the 1960s, he was asked if he were "really black" (p. 15).

> The identity of being a Black person was the identity of the Black movement . . . [W]e maintained the notion, the myth, the narrative, that we were really all the same. The notion of essential forms of identity is no longer tenable (p. 17). . . .
>
> We tell ourselves the stories of the parts of our roots in order to come into contact, creatively, with it. So this new kind of ethnicity—the emergent ethnicities—has a relationship to the past, but it is a relationship that is partly through memory, partly through narrative, one that has to be recovered. It is an act of cultural recovery (p. 19).

In this case, Hall recognizes "race" as a social construction that is linked with historical stories, which produce real consequences in the world and also change over time.

Butler (1990) theorizes that identities are maintained through *performance*. According to Butler, the rules that govern identity such as "gender hierarchy and compulsory heterosexuality, operate through repetition" (p. 145) of bodily gestures, clothing choices, social practices, and word choices enacted within certain times and spaces. This repeated reenactment of socially learned identities in various cultural venues is the mundane way that "normal" meanings are maintained.

Although we may perform our identities differently in various social locations, we usually imagine ourselves as positioned in relation to others. Images contribute to the ways we might see ourselves and feel comfortable acting. In every culture, men and women learn the historically preferred aesthetic codes, which are often gendered. Individuals may contest the dominant aesthetic or customs, but others who observe those codes will commonly condemn women or men who challenge the visual codes for their gender or position. For example, on June 22, 2005, an Iranian woman was shown on television challenging historical gender stereotypes by competing as race car driver in Iran.

Thus, this second approach suggests that images are learned within codes that refer to social identities that have been associated with histories, languages, and cultures. Individuals may decode those codes and use visual images to associate those histories with themselves or other people or challenge those identities.

Using the second approach, viewers might ask the following questions: With whom do I identify in the image? Do I identify more with people who appear similar to my age, gender, "race" category, sexual identity, economic background, ethnicity, religion, language, technology, or "modern" ways of life? Does this image remind me of experiences I have had or seen? What might my friends or family members say about this image? What might another person who identifies more directly with this image say about it? What ideas or feelings does this image make more, or less, possible to think or imagine for people of different social categories? Might someone be encouraged or hurt by the messages implied in this image? Why?

Approach 3—Sociocultural Historical Contextualization

This approach to knowledge construction is based on the work of Mirzoeff (1999) in art history, Foucault (1980) in philosophy, Wynter (1995), Berger and Luckmann (1966) in the sociology of knowledge, and Giroux (1999) in education. To engage images as visual culture, Mirzoeff (1999) has recommended historical research:

Visual Culture seeks to blend the historical perspective of art history and film studies with the case-specific, intellectually engaged approach characteristic of cultural studies . . . Visual culture, like any other means of sign analysis, must engage with historical research (pp. 12–14).

Ideas about historical research are inspired by the work of Foucault (1980) who has written:

> Let us give the term *genealogy* to the union of erudite knowledge and local memories which allows us to establish a historical knowledge of struggles and to make use of this knowledge tactically today (p. 83).

Images, like other languages, carry historically accented, power-bearing meanings, into particular historical, social, and political conditions. Rather than assume that any image means something outside the conditions of its production and social use, images function in discourse within histories.

The viewer might ask questions such as the following: Who produced this image? For whom and why was the art made? (If the image is a commercial product, who is the intended audience and what is the expected outcome?) What contexts may have influenced the production of this image, its distribution, and the current meanings associated with it? What was the artist's relationship with the people or things s/he is representing? What was going on in the world politically, economically, and socially at the time this image was made? What do you think a variety of people at the time might have said about this? Why? What social values or stories do you think the artist brought to this work? How do you think this image participates within other social or historical dialogues?

Approach 4—Intertextual Articulation

Scholars from various fields have been studying the interplay among historical documents, literature, art works, music, popular culture, or media events as sites of social interaction and meaning construction. They have proposed intertextual connections that have suggested ways that discourses of social and political meaning have been put into social use to extend or defend the interests of discursive communities. For example, Mirzoeff (1999, p. 129) has used transcultural and intertextual analysis to link nineteenth century images in art history with other "texts" such as novels, music, and popular culture to show how art has participated in cultural discourses to normalize and legitimate imperialism.

Intertextuality refers to the connection a viewer might make between any two "texts." Gramsci (1971) noticed that similar ideologies are located in many cultural, scholarly, political, economic, and popular "texts." Gramsci proposed that similar conceptions appear so often, in so many sources, that they appear to be "true," based on "common sense." Gramsci recommended linking popular cultural texts with other texts as a way to see a common ideology in various locations. How are texts connected? Hall (1986) theorizes about *articulation* as a connection.

> An articulation is thus the form of the connection that can make a unity of two different elements, under certain conditions. It is a linkage, which is not necessarily, determined, absolute and essential for all time. You have to ask, under what circumstances can a connection be forged or made? The "unity" which matters is the linkage between that articulated discourse and the social forces with which it can, under certain historical conditions, but need not necessarily, be connected (p.141).

Grossberg (1992, p. 54) has explained that "[a]rticulation links this practice to that effect, this text to that meaning, this meaning to that reality, this experience to those politics."

Using this fourth approach, viewers might ask, Can I link this image with any other cultural texts (such as music, books, toys, games, movies, environments, social practices, experiences, or historical events)? Why do I connect this image with these texts, memories, or associations? What elements, stories, values, and beliefs do these texts have, or do not have, in common?

Approach 5—Cultural Narratives / Discourses of Knowledge and Power

The Cultural Narratives approach is based on the work of Friedman (1998) and Hall (1986). Discourse is based on the work of Foucault (1977/1980), Fiske (1996) and Gee (1999), as discussed above. Discourse implies ways that language and other cultural texts are used to construct understandings about the world that open particular ways of thinking and acting, while disinclining other possibilities.

To address Approach 5 a viewer might ask questions such as the following: Does this image suggest beliefs or ways to think about groups of people, the land, animals, or the environment? Does the image you are viewing contribute to ways people define meaning or power that include some people and exclude others? Do you associate this image with cultural narratives that relate to meanings and power relations?

Approach 6—Potential Social Consequences

Mitchell has suggested that analyzing cultural forms as representations

> leads us to ask not merely what these forms "mean," but what they *do* in a network of social relations: who or what represents what to whom with what and where and why? Most important, it automatically raises the question of responsibility. (Mitchell, 1994, p. 423)

This approach is also based on the work of Freedman and Combs (1996) in counseling psychology.

Using this approach the viewer would ask questions such as the following: What effect might this image have in the world of images and ideas? What emotions, desires, pleasures or revulsions does this image evoke for me? If I were going to put the effects of this image into my life, how would I know it was there? If I were to step further into this way of thinking or being, how might it affect me or other people? Are there other larger social stories that team up with this image to influence ways that people think, feel, or act? What might I think, feel or do if I believed in the messages that I interpret from this image?

Approach 7—Response-Ability

In the previous section, Mitchell asks what responsibility we have to react or reply to the messages that visual images imply. Artists commonly make art to reflect their concerns. Many artists make art to challenge images and issues found in art and popular culture or by inventing images to express their views. Artists such as Faith Ringgold, Carmen Lomas Garza, and George Littlechild, have published children's books that offer images based on their own life experiences to challenge stereotypical images in popular culture.

Questions based on Approach 7 might include the following: How would you like to respond to the image you studied? If you could speak to that image, what would you say? Are the messages in this artwork manifest in your life? What alternative values are not suggested in this image? Can you imagine a future image that you could make to respond to the messages making it more or less powerful? What could you do that would strengthen or challenge the messages in this image in an artwork or in your life?

The following examples show how two preservice teachers have addressed the approaches mentioned above from their own perspectives. First, a Japanese-American woman interrogates the representation of a Japanese character, and second, a Euro-American man investigates images of *G.I.JOE®* and guns in terms of identify formation.

Images and Stories of a Japan-American Insider

Karen, who identified herself as Japanese-American, chose to study the cultural representations of a Japanese man, in the 1961 Paramount film *Breakfast at Tiffany's* in 1998.[1] Although the movie premiered in 1961, the video emerged in 1992 and has remained a popular classic in video stores.

In response to questions that correspond to Approach 1, Karen has identified codes of representation and expressed her connotations when she writes,

Holly's character is tall, slender and attractive. As her last name implies, Holly's actions are elegant, energetic and free. She would have been considered to be a mainstream person, by the way she talks, acts, and dresses . . . In contrast to Holly Golightly's character, Mr. Yunioshi is short, has a round face and bucteeth, has thick glasses and is unattractive. In every scene that Mr. Yunioshi appears in [*sic*] he is wearing a kimono, hachi maki (head band), and tabi (Japanese socks). These are articles of clothing that even in Japan are usually worn only during special traditional holidays. Even in Japan in 1958 when the book, *Breakfast at Tiffany's* was written, a Japanese person would not dress in the same way that Paramount Pictures portrayed Mr. Yunioshi. Also in contrast to Holly's graceful movements, Mr. Yunioshi's actions were highly exaggerated as being clumsy. I think that Paramount Pictures partially created Mr. Yunioshi's clumsy character to draw humor into the film. But I found his mannerisms to be offensive. His movements were stiff and robot-like . . . I believed that Paramount Pictures used Mr. Yunioshi's character to emphasize and contrast the character of Holly Golightly. By creating this contrast, the absurdity of Holly's fictitious life and dream life is more readily emphasized. (Karen, 1998)

Karen used codes such as the characters' physical characteristics, actions, speech patterns, clothing choices, and mannerisms to support her connotation that the Golightly character, played by Audrey Hepburn, would seem to be a more "attractive" and a "mainstream person," while the Yunioshi character, played by Mickey Rooney, was constructed to look "unattractive" and "offensive." Further, Karen argued that "the Japanese neighbor" was negatively developed to positively construct the "mainstream" character. Her argument was similar to that posed by Toni Morrison in *Playing in the Dark* (1992) about negative construction of Black characters used to positively contrast White characters in literature.

In response to questions based on her own subjectivities, Karen interpreted the character through her own lived experiences as a Japanese-American. Karen wrote that she felt "singled out as a viewer" since there are so few characters of Japanese ancestry in films in the United States. She also discussed the way she learned the meanings of the cultural symbols of clothing, bathing, tea drinking, and lanterns through her experiences.

Even watching this movie in the comforts of my own home, with my own sister made me feel embarrassed and self-conscious . . . The images that Paramount Pictures used to develop Mr. Yunioshi's character are all images that personally mean a lot to me. But in my life they are symbols of tradition, culture and respect. When I think of wearing a kimono, I think of Obon. Obon is a tradition that is carried out at my temple every year to respectfully say good-bye to loved ones who have passed away within the year. By dancing traditional folksongs on this day we are respectfully praying for the safe passage of our loved ones to a more serene place. On this occasion it would also be appropriate to see lanterns decorating this festive day (not in an apartment). (Karen, 1998)

In critiquing the representation of Mr. Yunioshi, Karen felt an "awareness of social, symbolic and cultural signification of body" (Moore, 1994: 43) that she transferred to herself.

Karen analyzed the use of language in the film within the context of the film text as well as the time during and after World War II, which follows Approach 3, Sociohistorical Contextualization.

[W]hen Mr. Yunioshi was talked about, he was referred to as the "dear little man" and the "little Jap." These references are derogatory because the emphasis of the adjective "little" can be seen as describing this character's stature and social status. The term "Jap" is also very derogatory. It is the discriminatory reference that originated around the time of World War II. This term has negative connotations of hatred and disrespect . . . Mr. Yunioshi was born in California and is Japanese American [but] he still speaks broken English. One line that is repeated throughout the movie by Mr. Yunioshi was "Miss Golightly, I must protest." Even though this line is grammatically correct, the wording is awkward and formal. This type of communication set Mr. Yunioshi's character apart from the other American characters. (Karen, 1998)

Karen connects the images, words, objects, actions, and mannerisms within the movie with other stories she has heard about her relatives' life experiences, which corresponds to Approach 4, Intertextual Articulation.

When I hear the word "Jap" I am reminded of the stories that my parents and grandparents have told me about their experiences around the time of World War II. Both sides of my family were stripped of all of their possessions and property and put into relocation camps. One camp was merely old horse stables at a deserted race track . . . Another camp that they were placed in were [*sic*] thrown together barracks that were not well insulated. During the years in which my grandparents lived in these camps, my dad and my uncles were born. The struggles and experiences that my relatives have experienced have really shaped who I am and what I believe in today. I get angry when people use the term, "Jap," so lightly without realizing or really understanding the meaning or the origin of the word. (Karen, 1998)

Karen located the images within the movie *Breakfast at Tiffany's* as participating with broader discourses of knowledge and power about the construction of Japanese-American identity, history, and culture, which is an example of Approach 5, Cultural Narratives / Discourses of Knowledge and Power. She juxtaposed her own family's history and the history of events during the World War II to show how her family were imagined as potential terrorists, while they were also citizens within their own country.

Karen realized the power of images and the potential social consequences, Approach 6, when she wrote,

> I have first hand knowledge of what the Japanese culture is like, but other viewers may not be equipped with the same information. In this case, I think it would have been more beneficial for the image of a Japanese character not to be seen at all rather than being misrepresented . . . Images in the media today need to better represent all in our society, including Asian Americans. Ethnically diverse characters not only should be present, but also should play non-stereotypical roles. (Karen, 1998)

Karen offered alternative ways to address representations of Japanese Americans in schools or other settings, which is an example of Approach 7, Response-Ability. Later in her paper, Karen discussed using a children's book, *So far from the sea*, about the Japanese internment, as well as films (such as *Rhapsody in June, Come see the paradise*, and *The Joy Luck Club*) that presented Japanese Americans or Asians in more diversified and dignified roles. She also stressed that many Japanese-Americans did not agree with the Japanese decision to bomb Pearl Harbor and that many non-Japanese-Americans did not agree with the internment. In fact, she reported that some White Americans moved into the camps to provide education and medical services for the Japanese American people and used their own money to buy supplies. These books, films, and historical information offer alternative images and narratives to construct diverse interpretations of this film and historical time.

In 2000, I assigned preservice teachers to write about one image that they would use as their central image and to construct a website that would engage their future students in discussions from many visual culture perspectives. In contrast to Karen's analysis of Japanese stereotypes in 1998, another preservice teacher in 2000 studied images of Japanese people taken by the famous photographer, Dorothea Lange, in the relocation camps and composed a website to help children understand the context that Karen described above.

G. I. JOE®—How Do Guns Shape Male Identification in Our Culture?

In 2000, Joel Blecha[2] selected an image of *G.I.JOE®* for his paper and website because of a tragedy that grabbed his attention.

> Just days after being assigned this project, a 6-year-old boy shot and killed his first grade classmate in a suburb of Flint, Michigan. I was

deeply affected by this tragedy since I am currently working with first graders. It got me thinking, "How much was this boy's behavior affected by the violent movies his father says he liked to watch so much?" Upon more thought, I realized how many violent and gun-related images young boys are exposed to. How do guns shape male identification in our culture? (Joel, 2000)

Joel described the comic book image that he selected using codes of representation.

This is the very first [cover] of the *G.I. Joe®* series for Marvel Comics® released in June of 1982. I searched with the keyword, *G.I. Joe®* on the [search] engine Dogpile and found [it on] the website, www.yojoe.com . . . The cover is an illustration of six members of the *G.I. Joe®* (five men and one woman) "Special Missions Force" in full action, their bodies poised amidst smoking explosions and their guns blazing in every direction . . . The cover is so incredibly busy that it is difficult to ascertain a point of emphasis. Initially my eyes were drawn to the massive tank in the background because of its dark blue coloration that works in contrast to the surrounding warmer reds, yellows, and oranges. The tank also brings unity to the image as it brings balance and represents the vanishing point of this cover's perspective . . . It appears that the reader is directly in the path of this attacking unit . . . I get the feeling that this juxtaposition of a "harmless attack" invites the reader to partake in this exciting adventure.

Joel reflects on the ways he learned to understand the meanings of *G.I. JOE®*.

I played with *G.I. Joe®* action figures and watched the animated television show throughout much of my childhood. My personal attachment to this image is therefore strong. This comic book cover is important for this assignment also due its depiction of random acts of violence. As I noted above, the way that the soldiers are firing their guns indiscriminately into the foreground is exemplary of the kind of casual or even "harmless" violence that *G.I. Joe®* purports.

When reflecting on questions regarding the sociohistorical contextualization, Joel wrote about the actual artists who produced this magazine cover, the corporation that sponsored them, and the layered marketing package of which this image was a part.

This image was created in the classic comic book fashion: pencil, then ink, and color. The penciler is Herb Trimpe. the inker is Bob McLeod with letterer, Jim Novak. Glynis Wein is the colorist and Larry Hama [is] the scripter . . .

Like many other children's toys, *G.I. Joes®* is a multi-dimensional product, spanning print, television, and toy manufacturing. This comic book cover is a way that the Hasbro® toy company could tie the television series into print media. These two mediums then were complimented by a huge toy campaign. Each year, Hasbro released a new series of action figures accompanied by attractive television commercials that showed young boys having a grand time playing with their new *G.I. Joes®* . . . The target audience for this image is young boys between the ages of 4 and 10. Guns and militaristic imagery are hallmarks of male gender typing. This is the theme of my educational web page: how guns shape male identification. Young boys seem to innately gravitate towards gun play and a desiring of firearm-based toys. *G.I. Joe®* caters to this tendency.

Joel considered the broader contexts within which the G.I. JOE package functioned in the United States and events in the world to which he linked the character.

The date of publication for this comic book is 1982. The United States and the Soviet Union were fiercely locked into the Cold War and President Ronald Reagan was making disastrous cuts to social welfare programs in order to feed the heightened arms race. Huge amounts of money were spent in the design and construction of such extravagant weaponry as the Stealth Bomber and the Star Wars orbital defense against nuclear attack. We are only now seeing how truly frivolous such spending was. The threat of Soviet missile attack was seriously inflated by our nation's intelligence agencies as Howard Zinn [1995] writes in his revealing book, *A People's history of the United States* . . ."In 1984, the CIA admitted that it had exaggerated Soviet military expenditures" [cited in Zinn, 1995, p. 571]. Zinn then writes how the people of the United States were conditioned to fear the Communists in the name of gross military spending. "[T]he creation of such a fear in the public mind was useful in arguing for the building of frightful and superfluous weapons. For instance, the Trident submarine, which was capable of firing hundreds of nuclear warheads, cost $1.5 billion. That $1.5 billion was enough to finance a five-year program of child immunization around the world against deadly diseases, and prevent five million deaths." [Cited Zinn, 1995, p. 571]

Joel interpreted the potential consequences of the G.I. JOE messages.

I believe that the vast popularity of Hasbro's *G.I. Joe* amongst young American boys is partially responsible for such propagandist efforts to rally the people around a united goal of a militarily strong state. I look at the cover of this *G.I. Joe* comic book and see it held by a little boy who would've thought a Trident submarine was perhaps the coolest thing imaginable. The toy version of this fearsome monstrosity would be at the top of his Christmas wish list.

In the course of his writing, Joel discusses many gendered and racial cultural narratives within which the G.I. JOE character and action figure participate.

The inclusion of the token female character has always baffled me, though. As we see on the *G.I. Joe* cover, one of these soldiers is the woman, Scarlett. I call her "token" because the only ways she was included in the plotlines was as the love interest or the damsel in distress . . .

As I sit at this keyboard and write, my eyes open wider and wider to the gender-stereotyping young children, including myself, were subjected to. It is done almost stealthily and insidiously, leaving me with feelings of confusion, curiosity, and bits of frustration, too. I now wonder about other attempts of such toy manufacturers to play to our children's subconscious . . .

The characters within this image are all white save for one African-American soldier named, Stalker. Even as a kid I wondered if this name was a bit unfair and perhaps prejudicial. The other characters have names that either describe[d] their specialization (Flash is the soldier who uses a flame-thrower and Clutch drives the vehicles) or make a cool connotation of some kind (Hawk and Rock 'n Roll) whereas Stalker just sounds a bit too sinister for a good-guy's name . . .

The only other minority group represented—to my recollection— is that of American Indians. The American Indian character on the *G.I. Joe®* force was always seen dressed in stereotypical garb of feathers and face painting. His name: Tracker.

Joel suggests the consequences of *G.I. JOE®* segments that show relatively little injury or death from warfare.

This is the same boy who grew up to be wowed by televised images of 'smart' bombs as they 'ingeniously' fell through air vents of Baghdad high-rises, blasting them as well as the people within into dust. By making warfare seem cool, exciting, and without consequences of serious injury or death (I recall the *G.I. Joe®* cartoon showing enemy pilots ejecting from their razed jets and parachuting to safety every time without fail), *G.I. Joe®* and merchandizing related to it work to desensitize an entire population against massive arms stockpiling and war in general.

Joel makes intertextual connections between the *G.I. JOE®* magazine cover and other cultural texts, such as Marine Corps advertisements.

> I connect fantastic images like this comic book cover to a current television commercial paid for by the United States Marine Corps. In a neo-medieval setting, a young man climbs atop a narrow bridge that spans a vast pool of fire surrounded by cheering spectators. Suddenly, an awesome lava giant emerges from the fire only to be destroyed by the sword-wielding man in fierce battle. The victor's plain garb then morphs into the uniform of a United States Marine as we hear, "the few, the proud, the Marines."

As Joel analyzes the images and texts, he thinks about the ways that G.I. JOE positions children to think about democracy and their role as citizens.

> A beige rectangular box in the lower right corner collectively describes the frenzied soldiers as, "THE ULTIMATE WEAPON OF DEMOCRACY!" What a grandiose and obviously loaded statement! How many young adults now unconsciously think that democracy is synonymous with warfare? In fact, President Bill Clinton's immoral bombing of Iraq has been named "Cruise Missile Diplomacy" by his White House staff. Our elected representatives (the cornerstones of a democratic republic) have a skewed idea of democracy if they attempt to achieve it through "diplomacy" via death and carnage.
>   These are the ideas that this *G.I. Joe®* image purports.
>   The parts I can decipher read: "INTRODUCING AMERICA'S NEW SPECIAL MISSIONS FORCE!" This bit of text makes me wonder if a young version of me once thought *G.I. Joe®* was an actual part of the American armed forces.

Although many preservice teachers have critiqued aspects of popular culture that they loved as children, they frequently expressed nostalgia about the images. I have found many preservice teachers have minimized analyses of the toys or have become angry because they felt personally attached when other students criticized toys they loved. In this case, Joel has admitted his nostalgia, yet he can also critique his favorite toy as a future teacher of young children.

> In closing I want to state that the above allegations of propaganda and stereotyping are merely the reactions of one individual to one image. These accusations come from an overall personal feeling of being misinformed as a young boy. G.I. Joe action figures were among my favorite toys and TV cartoon shows. Even amidst my critical analysis of this comic book cover, I find myself feeling nostalgic. I even remember smiling when I came upon the image on the Internet just a few weeks ago.

As an elementary educator, I need to hone my critical thinking skills. I must try to be ultra sensitive of mass media imagery for the child who is developmentally unable to be. Yet, at the same time, I should recall myself at that young age and remember how much I enjoyed playing with toy soldiers that came complete with an arsenal of firearms.

To respond to his concerns, Joel created a website that would help elementary students discuss violence. First, he asked the students to tell each other what they saw when they looked at the image of the G.I. JOE cover. Next he asked them to discuss memories of playing with guns. Second, he invited the students to look at art made by other children and himself as a child (as shown below). Third, he explored guns in popular culture by showing and discussing toy guns, posters, and lunch boxes of cowboys. He also compares the old cowboy narratives in film and television with the high definition, lifelike, graphics in video games, such as Duke Nukem and Doom. Joel wrote, "These two games in particular revolve around one character using an incredible arsenal of firearms to destroy level upon level of bad guys. Blood and gore are a common sight." Fourth, he offers questions and responses from diverse groups about questions of gun control. He provides links to three websites: one from the National Rifle Association (NRA), another "anti-NRA" site, and a third that contains "current legislation regarding gun control, what political candidates have to say on the subject, and interesting statistics." Finally, he included a learning plan in which he shows photographs of things that give him energy such as art, sports, and drama. He invites students to make an art work about a time when they felt energized without guns.

Figures 1–3 Drawings with magic marker. Fragments of one untitled drawing. Joel Blecha, six years old, 1984. Permission given by Joel Blecha, October 15, 2005.

On his website, Joel remembers the ways he would use drawing to illustrate his nightmares when he was four to six years old (see figures 1–3).

When I was 4 or 5-years-old I would get terrible nightmares. I remember going downstairs to tell Mom of these scary dreams. She would get me paper and crayons and ask me to illustrate the contents of my nightmares. What resulted was what many people would call *art therapy*. After drawing the monsters of my dreams, I would sketch their destruction with guns, bombs, and knives. Having successfully eradicated these creatures on paper, I could return to bed for a sound night's sleep.

Here are some drawings of mine that I produced when I was 6-years-old. (All three are portions of one larger drawing.) I can't remember exactly why I drew these depictions of war. I'm guessing they were greatly influenced by the *G.I. Joes®* that I was playing with at the time. Not only was I playing with the *G.I. Joe®* toys, but I was also watching the television show . . . Although these drawings are indeed violent, the only blood shed is from the poor seagull! Notice how in all three drawings the soldiers remain unscathed in the midst of gunfire. In the top two drawings, the soldiers' helmets protect them from being killed by bullets.

And in this bottom drawing, the pilot swims to safety after his warplane crashes afire into the sea. This lack of human casualty in my drawings is a hallmark of the G.I.Joe animated series. No soldiers from either side of the battle ever suffered an injury greater than a broken bone! Every time a Cobra jet was shot from the sky, the pilot ejected and gently floated away via parachute. (Joel Blecha, student teacher, field notes, 2000)

For the last five years Joel Blecha has been teaching first, second, and third grade students in Brooklyn, New York. I asked Joel to reflect on the paper and website that he wrote during his undergraduate preparation in light of his teaching experience. He responded,

In my five years of teaching in the New York City public school system, I frequently observe boys—whether it be first, second or third graders—using popsicle sticks, blocks, their fingers or any other oblong thing to represent guns. My response has always been the mantra: "No guns in school." I believe our proximity to the terrorist attacks of 9/11, in addition to the recent wars, have made me even more wary of kids' gun play. I see guns and tanks and warships—images similar to those in your book [chapter]—appear in their writing and drawings, too. I respond by encouraging my students to focus mostly on personal narratives during writing time. I realize that the war scenes I drew at age 6 were done at home, not at school. In addition, I was raised by two parents that taught me to value peace, not violence. I don't care to speculate whether there is a greater amount of violence in today's media, but I do know that my kids are exposed to a lot of it nearly every time they sit before video games and cartoons. I can't control that, yet I can work to create a classroom community of learners who favor fairness and peace over selfishness and hurt. (Joel Blecha, personal communication, October 11, 2005)

Conclusion

The work of these preservice teachers suggests that their interpretations of images are influenced by culturally learned codes of representation and their own subjectivities, learned as particular individuals through social interaction with people or cultural texts. Both preservice teachers provided social, cultural, or historical events to exemplify the context during which the images were produced. Both suggested the ways that these images and accompanying narratives participated within the networks of meanings and power to frame people as citizens of the United States and consumers of ideas and objects. Further, they discussed the consequences of the narratives in their own lives and imagined how other people might imagine and perform the narratives about Japanese people or military violence. Finally, they suggested alternative books or activities that would engage students in active investigation of narratives implied by their central image.

This chapter has focused on the ways that people link images with cultural narratives to construct meanings and perform those meanings in their lives. Seven approaches to culturally based image interpretation were recommended. These approaches were derived from scholars working in many disciplines whose scholarship contributes to Visual Culture Studies. The author examined visual images as cultural texts with preservice teachers as they explored their memories of forming image-narratives as children and their reflections as adults. Examples were given from narratives collected from preservice teachers concerning their childhood associations with images as cultural texts. This author recommends that teachers and young students investigate the multiple ways that images, as associated with cultural narratives, invite people to think, feel, act, and imagine themselves and others. She hopes that teachers may encourage students to challenge and possibly *transform* dominant associations between images and cultural narratives through various actions such as discussion, art making, writing, and drama.

Notes

1. All citations from Karen are from fieldnotes collected in 1998.
2. All citations for Joel in this chapter are taken from my fieldnotes in fall of 2000.

References

Butler, J. (1990). *Gender trouble*. New York: Routledge.

Berger, P., & Luckman, T. (1966). *The social construction of reality*. New York: Doubleday.

Ewen, S. (1988). *All consuming images: The politics of style in contemporary culture*. New York: Basic Books.

Fiske, J. (1996). *Media matters*. Minneapolis: University of Minnesota Press.

Foucault, M. (1980). *Power/knowledge. Selected interviews and other writings 1972–1977*. C. Gordon (Ed.). New York: Pantheon.

Freedman, J., & Combs, G. (1996). *Narrative therapy: The social construction of preferred realities*. New York: W.W. Norton.

Friedman, S. (1998). *Mappings: Feminism and the cultural geographies of encounter*. Princeton, NJ: Princeton University Press.

Gee, J. (1996). *Social linguistics and literacies: Ideologies in discourse*. London: Taylor and Francis.

Gee, J. (1999). *An introduction to discourse analysis*. New York: Routledge.

Giroux, H. (1994). Benneton's "world without borders": Buying social change. In C. Becker (Ed.), *The subversive imagination* (pp. 187–207). New York: Routledge.

Giroux, H. (1999). *The mouse that roared: Disney and the end of innocence*. Lanham, MD: Rowman and Littlefield.

Gramsci, A. (1971). *Selections from the prison notebooks*. New York: International.

Grossberg, L.(1992). *We gotta get out of this place: Popular conservatism and postmodern culture*. New York: Routledge.

Hall, S. (1986). On postmodernism and articulation. An interview with Stuart Hall. In D. Morley & K. Chen (Eds.), *Stuart Hall: Critical dialogues in cultural studies* (pp. 131–150). New York: Routledge.

Hall, S. (1989). Ethnicity: Identity and difference. *Radical America, 23* (4), 9–20.

Hall, S. (1997). The work of representation. In Stuart Hall (Ed.), *Representation: Cultural representations and signifying practices* (pp. 13–74). London: Sage Publications.

Hicks, L. (1992–1993). Designing nature: A process of cultural interpretation. Madison, WI: *Journal of multicultural and cross-cultural research in art education* (10/11), 73– 78.

hooks, b. (1994). *Teaching to transgress: Education as the practice of freedom*. New York: Routledge.

Jhally, S. (1987). *The codes of advertising: fetishism and the political economy of meaning in the consumer society*. New York: St. Martin's Press.

Kilbourne, J. (1999). *Deadly persuasion: Why women and girls must fight the addictive power of advertising*. New York: Free Press.

Lanier, V. (1982). *The perception of art. The visual arts and the elementary child*. New York: Teachers College Press.

Mirzoeff, N. (1999). *An introduction to visual culture*. New York: Routledge.

Mitchell, W.J.T. (1994). *Picture theory*. Chicago, IL: University of Chicago Press.

Moore, H. (1994). *A passion for difference*. Bloomington, IN: Indiana University Press.

Morrison, T. (1992). *Playing in the dark*. New York: Vintage.

Pauly, N. (2002). *Visual images linked to cultural narratives: Examining visual culture in teacher education*. (Doctorial Dissertation, University of Wisconsin, Madison, 2002). Dissertation Abstracts International (UMI No. #072699).

Pauly, N. (2003). Interpreting visual culture as cultural narratives in teacher education. *Studies in Art Education, 44* (3), 264–284.

Rogoff, B. (2003). *The cultural nature of human development*. Oxford: Oxford University Press.

Tabachnick, B. R. (1997). *The social context of teacher development*. Lisbon: Address to the School of Education at the University of Lisbon.

Wertsch, J. (1991). *Voices of the mind: A sociocultural approach to mediated action*. Cambridge, MA: Harvard University Press.

Wynter, S. (1995). 1492: A new world view. In V. L. Hyatt & R. Nettleford (Eds.), *Race, discourse, and the origin of the Americas: A new world view* (pp. 5–57). Was hington, DC: Smithsonian Institution Press.

Zinn, H. (1995). *A people's history of the United States 1492–present*. New York: HarperCollins.

C H A P T E R   7

The System of Reasoning the Child in Contemporary Japan

Jie Qi

The purpose of this chapter is to explore how various technologies have constructed the notion of childhood as a way of disciplining and self-disciplining in contemporary Japan. This is a study which attempts to disturb what has been taken for granted about children and children's nature. By applying Foucault's notion of governmentality (Foucault, 1977, 1984, 1986, 1991), I assert that, first, there are various technologies which have constructed disciplining and self-disciplining in contemporary Japanese schools and, second, that the relationships between the teacher and the child are multiple. Foucault's notion of power is that a multiplicity of actions engenders power, and power operates through discourse associated with the construction of knowledge. Moreover, Foucault's conception of governmentality allows us to rethink the relationships among self, other, and institutional discourse. Using Foucault's theories allows us to be suspicious about power. Each instance of power relations must be carefully analyzed, with the assumption that sometimes, many contradictory forms of power may be operating.

First, this chapter argues that the construction of the child in Japan, is not an ideological product but rather a discourse that involves a complexity of power relations. The relationship between the teacher and the child is a dual one. Moreover, the reasoning used to construct Japanese understandings of childhood is not simply controlled by the government through its sovereign power but is also shaped by multiple technologies.

Second, this chapter explores disciplinary power in Japanese schools and how it normalizes the child. What I consider the disciplining of the child is not such things as school regulations that legitimately restrain the child. Instead, I am referring to the multiple technologies that discursively construct the child. The use of positive language, for example, normalizes teachers as those whose praise. Through the use of these techniques to motivate children, Japanese schools normalize the idea that children need motivation in order to work, and that this motivating effect must be subject to a teacher's praise. Moreover, journal writing, cleaning classrooms, and group activities, as technologies of child management, discursively construct and normalize the "reasonable" child as hardworking, kind, cooperative, and having collective awareness.

Third, this chapter indicates that Japanese children are simultaneously governed and disciplined by teachers and are self-governing and self-disciplining. The purpose of disciplining children is to make children self-disciplined. By exercising self-discipline, children approach self- understanding, self-esteem, and self-actualization. The distinctive feature of self-discipline in Japanese schools tends to be self-classification, that is, one knows how to insert oneself into certain categories, such as the "problem child," "not self-motivated" or "lacking group awareness" according to the teacher. Once a child has been classified, he/she disciplines himself/herself to become "normal." Children who practice self-discipline must believe that it is for their own good. The process of practicing self-discipline is enjoyable; at least one is made to feel that it is enjoyable and fulfilling.

This chapter concludes with the assumption that what Japanese children are, and how they become what they are, are effects of multiple power relations. It urges readers to be skeptical about the discourses regarding "normal" or "reasonable" children. The categorization of normality or abnormality involves methods of inclusion and exclusion. It is hoped that this chapter will stimulate dialogue and debate about the interaction between power, culture, and the various constructions of the childhood.

The methodology of this study includes textual analyses and classroom observations. The primary texts are as follows: (1) the teacher guidelines issued by the Ministry of Education of Japan which includes *Gakusyushidoyoryo*, known in English as the *Course of Study*; (2) teacher's diaries, together with some other teacher's writings; and (3) in-service teacher training textbooks. In-service teacher training textbooks are issued by the Ministry of Education but are actually written by the school teachers and educational researchers. In Japan, in-service teacher training textbooks play an important role in training teachers. In this chapter, instead of discussing how in-service teacher training textbooks represent the ideology of the Ministry of Education, I explore the kind of pedagogical issues the in-service teacher training textbooks raise, and how they construct childhood in contemporary Japan. The teaching observation took place in a public elementary school over the course of three months when I worked as a specialist for immigrant children.

Multiple Disciplinary Technologies of
Normalizing Students

In this section, I explore the construction of teaching and learning in Japan not as an ideological product, but as a discourse that involves a complexity of power relations. Schooling is not simply under the control of governmental sovereign power but is shaped by multiple technologies and a variety of self-disciplinary techniques is a part of schooling. I consider the normalizing technologies of the child not through such things as school regulations that legitimately restrain the child but through multiple technologies that discursively construct the child.

The Usage of Positive Language

One normalizing technology is the use of praise. Praise, as used by teachers, is a dividing practice that creates dispositions and distinctions among students. Praise, as a normalizing technology, functions in teacher's construction of "good" or "bad" students.

Teachers are encouraged to use positive expressions to promote students' self-advancement. The Ministry of Education has instructed teachers with the following guidelines:

> Teachers recognize students individually, encourage them and often praise students for their improvement. Students, therefore, will have a feeling of satisfaction about their efforts and have a strong will to accept the challenge of the next task. (Monbusyou, 1985, p. 22; my translation)

The following was a simulated lesson scene in an elementary arithmetic class described in a teacher training textbook (Jinbo & Harano, 1982, pp. 160–161; my translation):

Teacher: What is 2 and 3? (Many students raised their hands. The teacher pointed to Taro.)
Taro: Seven.
Teacher: Really? Taro thinks seven is the result. Hanako, what do you think?
Hanako: Six.
Teacher: Well, Hanako thinks it is six. Is that so? Jiro, what do you think?
Jiro: Five!
Teacher: Certainly! The answer is five. Taro, Hanako, do you two understand why 2 and 3 is 5? Both of your answers are wrong, but neither of you hesitated to present your ideas. I like this kind of student.

This teacher praised rather than criticized students who got wrong answers. The use of positive words is supposed to promote students' attitude toward learning and self-advancement.

At the same time, the strategy of using positive expressions normalizes teachers as those whose praise will have a motivating effect on students, and it normalizes students as those who need motivation and whose motivation is subject to a teacher's praise. Teachers' praise is directed at changes in students' habits. Nevertheless, in this notion of praise, the body is disciplined as well as the mind. The present idea of praise brings into being certain dispositions related to how the student is to act and be.

The example of a teacher directing students to use positive words can be found in a teacher training textbook published by the Ministry of Education (Monbusyou, 1989, pp. 35–40). An elementary first grade teacher lets students encourage each other. When someone faces a difficult task, others say "*ganbare* (Never say die)" to cheer them on. Positive phrases, such as "good job," "nice student," and "kind child," fly past each other in the classroom. One student who was praised by others said: "I'm so happy. It makes me feel confident."

The Japanese word "*ganbare*," generally speaking, means "persevere" in English. According to *Webster*'s dictionary, the definition of persevere is to "continue steadily in doing something hard; sticking to a purpose or an aim; never giving up what one has set out to do." Literally, it can be translated to "Hold on!" "Bear up!" "Never say die!" "Keep it up!" "Keep at it!" and so on. All of these phrases mean: "whatever activity you are engaged in, do your best to the very end" (Duke, 1986, p. 123).

The spirit of *ganbare* is deeply embedded in the Japanese society. Duke (1986, p. 122) has summarized the idea below:

> Throughout the lifetime of the Japanese they are surrounded, encouraged, and motivated by the spirit of *ganbare*. It begins in the home. The school takes it up from the first day the child enters the classroom. It continues through graduation. The company then thrives on it. It engulfs every facet of society. It is employed in work, study, and even at play and leisure. *Ganbare* is integral to being Japanese.

It seems to be "natural" for teachers to use *ganbare*. Generally speaking, it is believed that "making children *ganbare*" means "building up children," "developing children's abilities" and "improving children," and that "making children *ganbare*" is for the children's' sake. Every child always has to *ganbare*, no matter whether one is at the top or the bottom, no matter what one is doing. If a child fails in an examination, the teacher would say "*ganbare!*" to encourage this student do better; whereas if a student does perfectly in an examination, the teacher would still say "*ganbare*" to keep the child working hard. The child has to *ganbare*: "do his best in whatever endeavor he has set before him. The tennis club, the mathematics lesson, the lunchtime duty, the *osoji* (clean classroom and schoolyard), the after school *yobiko* (preparatory school), and the entrance examination are all pursued with a sense of *ganbare*" (Duke, 1986, p. 143).

The use of *ganbare*, as both a technology of disciplinary power and a technology of the self, has discursively constructed what it means to be a "normal" child. According to Foucault (1988), technologies of disciplinary power determine the conduct of individuals and submit them to certain ends, whereas technologies of the self

> permit individuals to effect by their own means or with the help of others a certain number of operations on their own bodies and souls, thoughts, conduct, and way of being, so as to transform themselves in order to attain a certain state of happiness, purity, wisdom, perfection, or immortality (p.18).

This contact between the technologies of disciplining others and those of the self-disciplining is what Foucault called governmentality.

The use of *ganbare* itself is an unquestioned way of thinking and talking about the self in Japanese society. *Ganbare* is a governing practice, creating particular ways in which children learn to "think, and see themselves in the word" (Popkewitz & Brennan, 1997, p. 293). In this sense, *ganbare* is a disciplinary technology of the self. Through the continuous, repetitive use of teacher's *ganbare*, students internalize this way of thinking about themselves.

Moreover, such a spirit of perseverance tends to be exhibited within a group. In Japanese, there are many expressions which construct the conception of collective spirit, for example "*minna de ganbare* (let's work hard together)," "*minna tomodachi* (let's be friends)," "*minna de nakayoku* (let's get along well)" or "*minna de asobou* (let's play together)." Teachers use these phrases very frequently at the school. These phrases have discursively constructed the notion of group-oriented spirit, which I further discuss later in this chapter.

Schooling is what Popkewitz (1998) has described as "the production of the rules embodied in action and participation" (p. 39). Such production is a disciplinary power exercised through Japanese schools. This disciplinary power is linked with multiple technologies such as praise, the use of *ganbare*, group activities, and so on.

Normalizing Students: Through Writing Journals

Another technology, which influences the normalizing of the child, is writing journals, called "*Seikatsu Noto* (Daily Life Notebook)." Foucault (1980) has explained:

> Truth is a thing of this world: it is produced only by virtue of multiple forms of constraint. And it induces regular effects of power. Each society has its regime of truth, its "general politics" of truth: that is, the types of discourse which it accepts and makes function as true (p. 131).

Making children write journals is teachers' daily work as prescribed in the *Course of Study* and teacher training textbooks. Students write in the

Seikatsu Noto every day. The style of the *Seikatsu Noto* is described below
(Fukuzawa, 1996, p. 305):

> Spaces for school days were a record of homework assignments, with
> a place to check and review the day's work and a block to use as a
> diary of events and feelings. The space for Sunday was a time line
> which students color coded [to indicate] how they spent their day. In
> the small section at the bottom entitled "reflection on this week," stu-
> dents checked off answers to questions about how virtuous they had
> been: Had they done good deeds, helped at home, studied enough,
> done their home work and been healthy?

Virtuous here is coded as good deed, helping at home, studying enough,
finishing home work, and being healthy. Particular Japanese discourses
have constructed the notion of *virtuous* as being sound in mind and body.
Generally and socially speaking, it is believed in Japan that one will not be
sick if one has a strong mind and a fulfilling life. Sickness is considered
somewhat *non-virtuous*. Therefore, for a student, being healthy is as important as studying hard, finishing home work, or helping at home.

Students hand in the *Seikatsu Noto* to the teacher at various intervals
(usually once a week). After reading, the teacher writes down his or her
comments on the Daily Life notebook. The student's daily life becomes a
text that the teacher comments on and tries to modify. One teacher guideline issued by the Ministry of Education suggests how teachers can reform
the "non-virtuous" students by responding to their journals:

(1) (The teacher) lets the student look straight at his/her present con-
    duct, and helps the student figure out whether this conduct was
    compelled by others or caused by his or her own will.
(2) While trying not to control the student's free thought, (the teacher)
    stimulates the student to judge whether his/her present conduct is
    the best choice for himself/herself and the society.
(3) If the student realizes that his/her conduct is not desirable, (the
    teacher) suggests to the student that s/he makes a new conduct plan,
    and that it is important to make the plan achievable in order to have
    a successful experience.
(4) (The teacher) intently watches the student's action and encourages
    the student. (The teacher) praises the student for his/her improve-
    ment. When the student did not materialize his/her plan, (the
    teacher) does not blame or punish the student but advises him/her to
    make a new plan (Monbusyou, 1990, p. 46, my translation).

This strategy of how to save "non-virtuous" children provides teachers
with a normalizing management of children. What it means "virtuous" or
"non-virtuous" is politically, socially, and historically constructed.

However, in contemporary Japanese education, generally speaking, a "non-virtuous" child is individualistic and tends not to follow traditional norms. It is through writing a journal that children are discursively constructed and normalized by the coded value which is inscribed in the teacher's response. However, what it means to be "virtuous" is historically contingent and is embedded in the social context. Thus, journal writing is a construction of the self though technologies of disciplinary power and technologies of the self involved in multiple power relations. These power relations involve the use of *ganbare*, the incitement to develop a strong sense of group belongingness, praise, and journal writing.

Collective Activities

Generally speaking, Japanese students are famous for their collective behavior. They wear the same uniform, eat the same school meals, and act alike. For Japanese, this disposition tends to be praiseworthy, whereas many foreigners see it as enigmatic. What I am concerned about here is not to judge whether this behavior is good or bad but about what technologies effect the value of group orientation.

The sports day is held in every school playground in the spring and fall, respectively. The distinctive feature of the Japanese sports day is that the sports day is not for the sake of sports. For the school, it is to improve students' collective spirit. Usually, all the games are group games, especially at elementary or junior high school. Group games must include everyone in the class. Using "*minna de ganbare*" (Let's work hard together) to express the Japanese sports day is certainly an apt remark.

For example, rope skipping, relay and ball rolling are typical class versus class games within a grade. Students begin practicing many weeks before the sports field day. The motto is "*minna de ganbare*." All class members have to take part in group games. If one child, for instance, cannot master rope skipping, the others will encourage and help him/her. The child who cannot do rope skipping, as one member of the class, "*issyou-kenmei ganbare* (try as hard as s/he can)" in order not to obstruct the whole class. Trying together, sharing the same experience, enjoying the same happiness, all of these are addressed. The Japanese sports day can be seen as a strategy of normalizing students.

In Japan, cleaning classrooms and schoolyards is naturally understood as the students' job. The cleaning activities in Japanese schools are not only for the sake of hardening students' bodies but also for disciplining the students' souls. A teacher training textbook issued by the Ministry of Education demonstrates how an elementary school first year class' group consciousness was improved by cleaning classrooms (Monbusyou, 1989, pp. 96–101). Choosing a job and figuring out how to accomplish it in a group developed the idea that "the group belongs to all and everyone belongs to the group, which means each individual's work contributes to the entire group, while the group is the reason why each individual exists. In disciplinary power, the discourse of self confidence functions as a technology of group membership and self identification. It is believed that the task of cleaning is essential to the students' education and emotional well being and they regard this duty as 'a major part of education,' ways to teach students how to work with others and how to care for themselves" (LeTendre, 1996, p. 285).

A routine or ritual of morning school meeting is another strategy for developing collective consciousness. The Ministry of Education instructs schools and teachers that "[For students,] listening to the principal's and the head teacher's moral discourse in the morning school meeting is effective in cultivating the habit of collective activities such as paying attention to others' speech and forming ranks" (Monbusyou, 1985, pp. 23). Besides the morning whole school meeting, there are two short meetings (usually ten minutes) a day and one long meeting a week. The daily meetings are called "*asa no gakkyukai* (morning class meeting)" and "*owari no gakkyukai* (closing class meeting)." Students take turns being in charge of the meeting. Usually two students are responsible for one day. The two students work together to organize the meeting. In the morning class meeting, students talk about what activities they have to accomplish during that day, and in the closing class meeting, they summarize how activities, decided in the morning class meeting, have been done. The two students on duty usually praise good students and good deeds rather than criticize bad students for bad conduct in the closing class meeting.

The weekly meeting is held on Saturday (schools in Japan are six days a week) and is run by the class officer. The purpose of the weekly meeting is to summarize the students' work during the week and what all class members have to "*ganbare*" for the next week. Through such experiences, students learn how to work with a peer and how to organize the class. Students are reiterating the "official" discourse. It looks like students are in charge, but they act in ways that are "normal" and they embody a discursively constructed subjectivity. It is unthinkable for students to run the meeting in a critical or subversive way.

School lunch is considered a technology for improving collective consciousness, as well. Unlike some other countries, for example, America, where students have lunch in the cafeteria, Japanese elementary and junior high schools do not have cafeterias, and students are not allowed to bring their own lunch to the school. Students have to eat lunch in the classroom. The school lunch is supplied by the school. School lunch began soon after the World War II for the purpose of nutrition. However, in the present day, providing school lunch is not for the sake of nutrition anymore but for disciplining students.

Every day, there are four or five students who are on duty. Their jobs are to bring lunch to the classroom, serve the meal to all students, get together tableware and bring them back to the kitchen. All students sit in a fixed group to eat lunch. The time when students should start to eat lunch is when the students who are on duty make a sign and all students say "*Itadaki masu* (Let's start)" all at once; and the time to finish lunch is when the students who are on duty give a signal and all students say "*Gohiso sama desita* (Thank you for the meal)" all at once again. No one is allowed to eat before or after. Leaving a dish unfinished is prohibited. Talking and joking during the lunch time are prohibited.

In the United States lunch time is a "break" from the pressures of the classroom. Students are very much free to talk, eat and joke as they please. Their lunchroom behavior is not supposed to be like their classroom behavior. Eating is personal—not part of schooling. Unlike the United States, in Japanese schools, not even eating is exempt from the demands of public convention. The United States school system includes "breaks" for lunch and recess and acknowledges classroom demeanor is somewhat artificial (you only have to act that way in the classroom). However, the Japanese system gives the impression that classroom discipline is "real life." It is the way all parts of life (including eating lunch) ought to be lived. School lunch functions as a disciplinary technology to train students not only in proper table manners but also in collective consciousness.

Group activities are often recommended by the Ministry of Education. The following example given by the Ministry of Education has shown how group activities changed a student's group awareness (Monbusyou, 1989, pp. 108–113).

There was an elementary third grade male student who was considered to lack a collective spirit. He often joked around during group activities, was irresponsible about his job, and took no interest in class events when he thought he had nothing to do with those events. The teacher conducted a robot-making event whose purpose was to increase this student's group awareness since "holding down, tying up and pasting, all this work needs other people's help" (Monbusyou, 1989, p. 111, my translation). The male student worked with some other students. "While making the robot, he not only had fun but also recognized that he belonged to the group" (Monbusyou, p. 113, my translation).

Language, journal writing, cleaning, meeting and school lunch, as the technologies of student management, are only a few examples that the Ministry of Education instructs schools and teachers to apply. These technologies discursively construct and normalize "reasonable" students as hard-working, kind, cooperative, and having collective awareness. Such reasonableness has become Japanese students' nature and this nature has been often taken for granted. There is nothing natural, necessary, or inevitable about the present Japanese students' nature. The value as to what is considered reasonable, of course, is socially constructed and historically contingent.

The Student as a Self-disciplined Subject

Foucault's conception of power have provided for a complexity of power relations. In the governmental state, individuals are not only the target or object of government. The relationship between governing and the governed is a complex one. Individuals are simultaneously governed subjects and take part in their own governing. Individuals become "the correlate and instrument" of the governmental state (Burchell, 1991, p. 127). On the one hand, the state outlines a possible art of government that depends on numerous techniques for disciplining individuals to be rational citizens. Taking care of one's own health, hygiene, and education, for example, are responsibilities of reasoning citizens. Individuals have shifted from rationally governed subjects to spontaneous problem-solvers, that is, they practice the appropriate forms of "technologies of the self."

The Ministry of Education has elaborated strategies for teachers: "Disciplining students is for the purpose of making students self-disciplined. By exercising self-discipline, students approach self-understanding, self-esteem and self-actualization" (Monbusyou, 1982, p. 44, my translation). This type of power relation is a complicated one. It is the opposite of the traditional sovereign power issue that assumes that power comes from the top and that power is employed by the sovereign to rein in the populace. The new type of power is of unknown origin. Power only exists when it is exercised, and power circulates.

Self-discipline requires self-motivation. The following story appeared in the teacher training textbook issued by the Ministry of Education. It is about an elementary fifth grade teacher working with a male student who had poor grades in all subjects (Monbusyou, 1989, pp. 89–95). The teacher concluded that a lack of self-motivation and self-esteem were the main reasons which prevented him from studying. In a mathematics class, the teacher wrote questions on the board and let students solve them individually. Seeing that the male student had difficulty solving the problem, the teacher softly put a hint card on his table. Finally, the student solved the problem. The teacher let him present his result to the whole class. It made him feel very honored. After that, the male student started to work hard on mathematics not only at school but also at home. He became enthusiastic about his study and "enjoyed" working hard. By the end of the semester, he had improved in all his studies.

The relationship between discipline and self-discipline is an intricate one. Discipline is to train the students to practice self-discipline for their own good. The technologies of discipline try to make students be enthusiastic about the task, enjoy the working process, and finally acquire self-esteem. By this means, this practice of self-discipline is engaged in the actualization of discipline.

The usage of *ganbare*, as we have discussed earlier in this chapter, is a technology of disciplinary power. Simultaneously, the purpose of using the word *ganbare* is to discipline students to exercise self-discipline. The teacher "naturally" uses *ganbare* and the student accepts *ganbare* as a "natural" part of his/her life. The student is to *ganbare* about everything at home and at school. The student is "naturally" inculcated with *ganbarism* and s/he believes that *ganbare* is for his/her own good.

In Japanese schools, teachers are advised to involve students in *gakkyu-zukuri* (creating classhood). Students "talk about what kind of class they want to be and what chores, promises and goals they will need to become that kind of class" (Lewis, 1996, p. 86). *Asa no gakkyukai* (morning class meeting), *owari no gakkyukai* (closing class meeting) and weekly meeting are opportunities for students to exercise power. Students, in turn, take charge of meetings. *Owari no gakkyukai* is also called *ichinichi no hanseikai* (critical self-reflection meeting of the day) and the weekly meeting is also called *syumatsu hanseikai* (critical self-reflection meeting of the week). During these meetings, students ask themselves: "Have I done anything good? Have I been mischievous? Have I been nice to classmates? Have I tried my best to help classmates? How can I do better? Is there anything I can do in order to make the class better?" Such self-reflection leads students to choose goals for self-improvement.

The Ministry of Education's teacher guidelines have shown how such self-discipline can be exercised by the students (Monbusyou, 1989, pp. 54–60). An elementary second grade homeroom teacher struggled to change students' attitudes to the whole class and the lessons. The teacher let students list what kinds of things had happened in the class, which made them uncomfortable, and furthermore discussed how to prevent them from happening again. The students "spontaneously" decided on the following: (1) making a carp streamer and putting it up in the classroom; (2) making strips on which to write rules and hang the strips on the carp streamer; (3) putting one strip into the carp when the rule has been followed by all class members; (4) taking out the strip from the carp if anyone broke the rule; and (5) ensuring that the rules written on the strips are decided by the entire class.

The following five rules were made: (1) do not exclude anyone from group games; (2) share the class ball and long skipping rope with others; (3) sit down in the chairs and wait until the bell rings for class; (4) do not laugh when someone's answer is wrong; and (5) prepare for the next lesson during the break. The first item was followed and the corresponding strip was put into the carp streamer in the first week. Students were looking forward to seeing the swollen carp streamer. By the end of the semester, all strips were put into the carp streamer. Putting strips into the carp streamer, as the teacher concluded, is instilling the rules into students' souls. Although the rules became invisible, they were deeply engraved in the students' minds. Thus, the students practiced self-discipline according to the teacher's expectations, but in a set-up that made it seem as though the students had chosen the rules themselves.

The distinctive feature of self-discipline in Japanese schools tends to be self-classification, that is, one knows how to apply oneself to classified categories, such as the "problem student," "not self-motivated" or "lacking group awareness" according to the advisory teacher. The technologies of discipline that construct a particular identity related to group membership. The classified individual disciplines himself/herself to be a normal or natural student in "normal" classroom discourse. People who practice self-discipline must believe that it is for their own good. The process of practicing self-discipline makes the student feel satisfied with himself/herself.

Conclusion

One may wonder if there is no absolute freedom and autonomy in modern society, if the modern individual practices self-discipline and this self-discipline is exercised through desires and pleasures. As we have explored in this chapter, the discourses embodied in the in-service teachers' training textbooks issued by the Ministry of Education are involved in a complexity of power relations, for example, discipline and self-discipline. The disciplinary power found in teacher training textbooks seen as technologies are praise, journal writing, and group activities such as cleaning that discursively construct the child.

On the other hand, Japanese children are involved in the practice of self-disciplining. Disciplining students is for the purpose of making students self-disciplined. Generally speaking, the teacher training textbooks have discursively constructed "normal" childhood in Japanese schools as being able to exercise self-discipline, to approach self-understanding, to develop self-esteem and to bring about self-actualization. Teacher activities such as *ganbare*, group activities (e.g., meetings, cleaning classroom, school lunch), journal writing, and praise are all found to be governing practices that create particular children's dispositions. These activities are disciplinary technologies of the self. Japanese children internalize particular ways of thinking about the self with teacher-directed activities. The distinctive feature of self-discipline in Japanese schools tends to be self-classification, that is, one knows how to apply oneself to the tacit classified categories, such as the "leadership student," "average student" or "lacking group awareness." The technologies of discipline construct particular identity related to group membership. Then, the classified individual disciplines himself/herself to be the "normal" or "natural" child in the "normal" and "natural" classroom discourse.

The use of *ganbare* itself is an unquestioned way of thinking and talking about self in Japanese society. *Ganbare* is a governing practice, creating particular ways in which students learn to perceive themselves. In this sense, *ganbare* is a disciplinary technology of the self. Through the continuous, repetitive use of teacher's *ganbare*, children internalize this way of thinking about themselves. Moreover, such spirit of perseverance tends to be exhibited within a group. In Japanese, there are many expressions that construct the conception of collective spirit.

The consideration of Foucault's theory in this study was a political strategy. I use Foucault as one of multiple poststructural theorists to open up new possibilities for rethinking systems of reasoning related to schooling practices in Japan. It is my hope that this study will usher in new winds to the field of educational research in Japan and also will inspire educational researchers to frame critiques in ways that were previously "unthinkable." Power only exists when it is exercised and circulates. Foucaultian scholars cannot predict in advance where power will work in the future and what power might do or not do. Neither is it possible for researchers to foretell which actions might be forms of "resistance" and which might be "compliance." Each instance of power relations must be carefully analyzed.

Foucault's notion of power enabled me to focus on the construction of child nature in Japanese educational discourses involving convoluted networks of power relations. Governmentality, as a technology of disciplinary power, is a helpful tool in analyzing ways power circulates in Japanese schools. Moreover, in employing Foucault's notion of power to rethink the production of reason in schooling practices, we can read educational discourse differently. Concepts such as governmentality and power/knowledge allowed me to open up new spaces in conceptualizing how the production of reason occurred.

References

Burchell, G. (1991). Peculiar interests: Civil society and governing the system of natural liberty. In G. Burchell, C. Gordon, & P. Miller (Eds.), *The Foucault effect: Studies in governmentality* (pp. 119–150). Chicago, IL: The University of Chicago Press.

Duke, B. (1986). *The Japanese school: Lessons for industrial America.* New York: Praeger.

Foucault, M. (1977). *Discipline and punish.* (A. Sheridan, Trans.). New York: Pantheon Books.

Foucault, M. (1980). *Power/Knowledge: selected interviews and other writings, 1972–1977.* New York: Pantheon Books.

Foucault, M. (1984). Truth and power. In P. Rabinow (Ed.). *The Foucault reader* (pp. 51–75). New York: Pantheon Books.

Foucault, M. (1986). *Disciplinary power and subjection.* In S. Lukes (Ed.). *Power* (pp. 229–242). Oxford: Basil Blackwell.

Foucault, M. (1988). Technologies of the self. In M. Martin, H. Gutman, & P. Hutton (Eds.), *Technologies of the self* (pp. 16–49). Amherst, MA: The University of Massachusetts Press.

Foucault, M. (1991). Governmentality. In G. Burchell, C. Gordon, & P. Miller (Eds.). *The Foucault Effect: Studies in governmentality* (pp. 87–104). Chicago, IL: The University of Chicago Press.

Fukuzawa, R. (1996). The path to adulthood according to Japanese middle schools. In T. Rohlen & G. LeTendre (Eds.), *Teaching and learning in Japan* (pp. 295–320). Cambridge: Cambridge University Press. Jinbo, S., & Harano, K. (1982). *Mondai koudo ni torikumu seitoshidou.* Tokyo: Gyosei.

LeTendre, G. (1996). Shido: the concept of guidance. In T. Rohlen & G. LeTendre (Eds.), *Teaching and leaning in Japan* (pp. 275–294). Cambridge: Cambridge University Press.

Lewis, C. (1996). Fostering social and intellectual development: the roots of Japanese educational success. In T. Rohlen & G. LeTendre (Eds.), *Teaching and leaning in Japan* (pp. 79–97). Cambridge: Cambridge University Press.

Monbusyou (1982). *Kyoikukatei to seidosidou: Kotogakkohen.* Tokyo: Okurasyo Press.

Monbusyou (1985). *Tyugakko niokeru kihontekiseikatsuyukan no sido: Situke no teityaku o hakaru.* Tokyo: Okurasyo Press.

Monbusyou (1989). *Seitosido o meguru gakkyukeieijyo no syomondai*. Tokyo: Okurasyo Press.

Monbusyou (1990). *Gakko niokeru kyoikusoudan no kanngaekata & susumekata: Tyugakko & kouto-gakkouhen*. Tokyo: Okurasyo Jinbo, (1982) Press.

Popkewitz, T. (1998). *Struggling for the Soul: The politics of schooling and the construction of the teacher*. New York: Teachers College Press.

Popkewitz, T., & Brennan, M. (1997). Restructuring of social and political theory in education: Foucault and a social epistemology of school practices. *Educational Theory, 47* (3), 287–313.

CHAPTER 8

The Specter of the Abnormal Haunts the Child: A Historical Study of the "Problem-child" in Brazilian Educational Discourses

ANA LAURA GODINHO LIMA

When one looks at current Brazilian education discourses about behavioral problems in schools, it seems like most of the concerns with badly behaved children could be explained by the conditions of life in Western families in current times. If a child is aggressive, it is because his/her parents are divorcing; if a pupil lies, he is trying to get more attention from a workaholic mother; if another one seems frightened, it is the effect of too much time in front of television watching violent cartoons or playing videogames and so on. Discourses about the influence of the home's emotional atmosphere on the child's conduct, as well as claims that the teacher should investigate what is going on in the pupil's life outside of school in order to understand them are taken to be new and progressive. However, statements that create a cause/effect relationship between children's habits and relationships in the family and misbehavior at school are one of the products of a historically bounded system of reasoning through which ideas of emotional development and social adjustment are constructed. These discourses ignore that since the first half of the twentieth century specialists have explained behavioral problems as caused by psychological factors relating to family dynamics.

In this chapter, I seek to describe attempts carried out since the late 1920s by diverse specialists to comprehend and deal with "problem-children." This concept was created to designate those considered defective in their emotional development, conduct, or social adjustment because of an unsuitable family or social environment. The category "problem-children" blurred the distinction between "normal" and "abnormal." Although they were deviant, they could not be considered to be really "abnormal" because their deficits were not the result of a biological defect but merely a product of inadequate education. This definition of the "problem-child" had two important consequences. The first one is related to treatment. While there was little hope for a cure in the case of abnormal children, it was possible to recuperate the "problem-child" through pedagogical measures. The second consequence relates to the scope of the category. Once it was possible to link behavioral problems to the child's inappropriate living conditions, all pupils had the potential to become "problem-children" as their families could not be completely secure against such difficulties as unemployment, death of close family members, and divorce.

In the first section of the chapter, I present the concerns about "problem-children" as an issue of government in a Foucauldian sense. This is followed by considerations on how different overlapping discourses came together to produce knowledges of "problem-children" during the first half of the twentieth century and to propose ways of dealing with them in schools and through families. In the last sections, I return to present discourses. The expression "problem-child" can still be found in Brazilian education discourses. Although the causes of these problems are generally thought to be the same, the recommendations on how to deal with behavioral problems are different. As an example, while educators writing between 1920 and 1940 recommended that "problem-children" should be educated in separate classrooms, today they are unanimous in avoiding this type of exclusion. Instead, they argue that "problem-children," in addition to attending regular classes, need to receive individual treatment by psychotherapists, "psychopedagogues," or other specialists, depending upon the child's particular problem or diagnosis. Moreover, once it was possible for every child to become a "problem-child," with emotional or social problems because of difficulties at home or school, psychologists and educators thought it necessary to encourage children to talk about their feelings in school as a way to solve conflicts and to prevent maladjustment and school failure.

The "Problem-child" as a Government Issue

The category of "problem-children" emerged in Brazilian education discourses during the 1930s as specialists worked to comprehend and solve difficulties presented by children in regarding their conduct, their social relationships, and their learning. It is possible to state that in fundamental ways, Brazilian authors followed the ideas of their American and European peers, trying to adapt their concepts and explanations to the national context. It was common, for example, for articles originally released in international periodicals, such as the *School Journal and Educator* and the *Bulletin of the Pan-American Union* of Washington, to be translated and published in Brazilian journals on education. Additionally, most Brazilian educators writing about the "problem-child" quoted foreign authors in order to bolster their arguments.

Students' problems were thought to require an intervention into their attitudes and behaviors to obtain specific desirable results such as responsibility, motivation, sincerity, effort, autonomy, solidarity, respect for older people and colleagues, and so on. In this sense, Foucault's concept of governmentality is appropriate, since he refers to government as "the conduct of conduct" or as an "action over an action."

The word "conduct" has a double meaning that helps to clarify Foucault's notion of power. Conduct can mean to *lead others to act in a specific way, under more or less coercive methods*. Conduct can also mean *the control of one's own aptitudes in a space of relatively open possibilities*. As Foucault argues "The exercise of power consists in guiding the possibility of conduct and putting in order the possible outcome. Basically power is less a confrontation between two adversaries or the linking of one to the other than a question of government" (1983, p. 221). Foucault attributes to government the broader meaning which it had during the sixteenth century, when this term did not refer only to the State, but to diverse ways of organizing the actions of various groups: children, families, ill people, and souls:

> To govern, in this sense, is to structure the possible field of action of others. The relationship proper to power would not therefore be sought on the side of violence or of struggle, nor on that of voluntary linking (all of which can, at best, only be instruments of power), but rather in the area of the singular mode of action, neither warlike nor juridical, which is government. (1983, p. 221)

Considering power relations as government that people exert one over the other, with the aim to determine or modify the actions one over the other, Foucault (1983) highlights the importance of freedom in these interactions, which occur as in a strategic game, in which one action may be reacted to in multiple, even unpredictable ways. For Foucault, it only makes sense to speak about power when it refers to a relationship between free subjects, even if the space of freedom for some individuals is extremely narrow:

> Power is exercised only over free subjects, and only insofar as they are free. By this we mean individual or collective subjects who are faced with a field of possibilities in which several ways of behaving, several actions and diverse comportments may be realized. Where the determining factors saturate the whole, there is no relationship of power; slavery is not a power relationship when man is in chains. (In this case it is a question of a physical relationship of constraint.) Consequently there is no face to face confrontation of power and freedom which is mutually exclusive (freedom disappears everywhere power is exercised), but a much more complicated interplay (p. 221).

Considering the education of the "problem-child" as an issue of governance allows us to comprehend how it was possible to conciliate discourses privileging the importance of knowledge of children as individuals and respect for their natural tendencies with disciplinarian practices in school. The multiple interventions formulated by the educators since 1930 to solve the behavior problems that children presented in school were not intended to restrain pupils' freedom. On the contrary, interventions were proposed as ways to promote individualities by removing emotional or other types of obstacles that obstructed its free manifestation. A child was considered well adjusted when he or she was able to conduct himself/herself as an autonomous person in the school, a child who knew how to behave in a space of regulated freedom. At the same time, according to Progressive School thought,[1] the good school was one in which pupils had the opportunity to express themselves, to discover and realize their own potentialities, and in which teachers were prepared to meet children's individual needs. It is necessary to remember, however, that children construct identity in the context of their relationships with others in the school. Relationships were not only established with close colleagues and teachers but also with populational norms. A child's relationship with the population norm was set in comparison to the population in which the child was inserted. Each exam grade fixed the individual's position in relation to others, inscribing his condition as normal, subnormal, or supernormal. In this sense, knowledge of the individual was possible only when in comparison of individual scores to fixed patterns established for the whole population. This was true for the educational institution as well as for individual children, who discovered themselves through processes of socialization in the school. In this sense, it is possible to say that pupils were haunted by the specter of abnormality. Once identity was located somewhere in the normal curve, each new evaluation carried the risk of slipping into the realm of abnormality.

The Construction of Children's Souls

This section focuses on how children's souls should be constructed and on which difficulties could be found in this process according to educators and other specialists. During the 1930 and 1940 many overlapping discourses came together to produce knowledge of children: not only the Progressive School, but hygiene, psychology, and psychoanalysis offered their concepts and explanations to illuminate the challenging issue of the "problem-children."

Among the principles supported by the Progressive School movement was that the activity and the autonomy of pupils were indispensable conditions for the education of the modern individual. The individual child was to construct his own subjectivity and acquire self-control though relations with others, starting in the family and in the school. This experience of the self, through which the subject was constructed, was determined by a series of techniques that codified the kind of relationship that the individual established to himself and to others. The "problem-child," was to be dealt with through techniques promoting the child's capacity to notice his/her difficulties and his/her disposition to the treatment imposed on him by educators. Frequently, the problems were diagnosed as a lack of capacity to conduct themselves appropriately and to distinguish between reality and fantasy, true and false, fair and unfair. To help the child overcome his/her problems was, in part, to promote the capacity to self-evaluate and to correct themselves. With this purpose, educators recommended a series of techniques, such as procedures for leading the child to confess their guilt, to explain the reasons why he or she had done something dishonest, and to imagine themselves being in the place of the offended person. The educator's role, as the responsible intermediary between the child and their conscience was considered crucial in this process, as the educator helped them see their own mistakes.

Foucault (1988) describes four techniques that individuals deploy to understand their selves: technologies of production, "which permit us to produce, transform or manipulate things"; technologies of sign systems, "which permit us to use signs, meanings, symbols, or signification"; technologies of power, "which determine the conduct of individuals and submit them to certain ends or domination, an objectivizing of the subject" and technologies of the self

> which permit individuals to effect by their own means or with the
> help of others a certain number of operations on their own bodies and
> souls, thoughts, conduct, and way of being, so as to transform them-
> selves in order to attain a certain state of happiness, purity, wisdom,
> perfection, or immortality (p. 18).

Further, Foucault draws attention to the significance of relationships with others when he states that a master's help is essential to teach a person to take care of the self. The modern school directs activities based on the same ideas, in that they teach children techniques of self-knowledge, self-control, and self-transformation. Since this implies a "truth game," where someone more experienced dedicates himself to teaching knowledge and abilities to someone less experienced, power is inevitable in the relation between teacher and pupil, although it is not necessarily bad. However, as other scholars have pointed out many times, the power exerted in educational institutions may be perverse when it turns into arbitrary domination of the pupil by school authorities.

The Brazilian educational discourses of the 1930s and 1940s discussed the diverse type of beings the term "problem–children" referred to. The "problem–child" could be, among others, a boy or girl, rich or poor, a toddler or early adolescent, or a city or country dweller. To different types of beings, distinct patterns of normality were attributed and diverse therapies were suggested. Generally, however, the analysis of the discourses indicates that those children more frequently considered "problems" were those who were more introverted, who had a rich interior and were distracted by their own fantasies and, consequently distanced from the "real world," and those who lied because they did not separate their desires and dreams from reality. After the introduction of psychoanalytical theoretical frameworks into the educational field, those who demonstrated an auto-aggressive potential, revealing a desire for self-punishment were also pathologized. Educators were also concerned with those children who were aggressive or with those who had a tendency to dominate others as well as the shy, passive child who resisted participating in the group activities. In the field of family relationships, the child who was over-loved was problematic as was the rejected child.

The Administration of Student Populations

In the final chapter of *The history of sexuality I: An introduction*, Foucault (1979) refers to relations between the State, the individual, and the population and their relation to issues of life and death. In this text, Foucault discusses a transformation that occurred in sovereign power, which until the seventeenth century emphasized the power to kill but now focused on the responsibility to conserve the life of the population. This power over life was concretized across two poles: first, the investment in the individual body by the diffusion of disciplinary techniques in the whole social body, a process that started in the seventeenth century. The aim of such techniques was simultaneously to expand the strength and the subjection of the man-machine, by means of a detailed control exerted over each part of his body. Second, since the middle of eighteenth century, government dedicated itself to know and intervene in the biological processes of the population: birth and death rates, health and illness, the duration of human life and the factors which determined such things were studied and administered by biopolitics. The emergence of biopolitics and its concern with life created a need for corrective and regulative mechanisms to organize individuals according to criteria of normality instead of simply imposing death. This condition changed the status of law, which began to exert a normalizing function.

Thus, biopolitics distributes people according to certain norms and aims to protect the life of a population. It functions to construct differences within the population and identifies groups contributing to the development of the population. It also identifies threats to the population among those presenting a hazard to health, the economy, freedom or to the life of the population, such as criminals and dangerous classes, the feeble-minded and the imbeciles, the degenerate and the unemployable. Once identified, authorities can create initiatives to prevent their appearance, to regulate them or even to eliminate them.

The notion of biopolitics, then, is valuable for understanding the thoughts and recommendations designed to protect, prevent, select, and correct children who are considered problematic. It was not only an issue of treating local individuals who were considered resistant to education, but also to study the entire child-aged population in a way that made it possible to identify the groups that could become problems in the school and to act upon them in a preventive way. In addition, intervening in every process that shaped children's lives seemed prudent because not intervening could lead to difficulties in the schools. In this sense, it was the child's relation with his context or his habitat that needed to be governed: the social and economical conditions that determined the quality of nutrition, habits of hygiene, routine, free time, relationship with parents and siblings, all were significant elements in the administration of the school population.

The book *The ABC tests*, by Lourenço Filho (1933/1974) that I analyze later in this chapter offers good examples of this kind of concern typical of biopolitics. Filho developed an instrument that tried to act upon the difficulties of certain groups of children to learn how to read and write, even before such difficulties appeared. Educators took many aspects related to pupils' life into consideration in the search for the causes for immaturity that some children displayed. Also, the notion of biopolitics allows me to situate the "problem-child" in relation to the broader set of preoccupations of government with the progress of society.

Childhood as a Social Problem

The vast majority, however, we could say the 90 percent of children seen as "abnormal" are in fact difficult children, "problems," victims of a series of adverse circumstances, that we are going to analyze in this book, and among which the maladjustment conditions of the social and familiar environment need to be highlighted (Ramos, 1939, p. 11).

Bearing in mind the definition of the "problem-child" proposed by the Brazilian pediatrician and educator Arthur Ramos in the quotation above, it is necessary to clarify the meanings of expressions like "social maladjustment" and "familiar maladjustment," which were frequent in the educator's discourses in the 1930s and 1940s.

Nikolas Rose (1999) states that while the nineteenth century saw the appearance of the calculable and classifiable individual and the rise of diverse techniques of individualization and comparison in relation to a norm, by the middle of the twentieth century it was possible to verify the appearance of the "social" individual. This individual was constituted through diverse social influences in which adjustment and happiness depended on the quality of his relations with the group in which he lived. The causes of delinquency and work problems were no longer only viewed as individual or family problems, but instead, were now viewed in relation to the individual environment or habitant. Courts of law dealing with minors began to consider delinquents more as victims of a maladjusted social environment than as guilty of infractions they had committed and instituted therapeutic penal regimes. In addition, children's difficulties in the school were now considered as consequences of an inadequate family environment. As Popkewitz and Bloch (2001) note,

> Increasingly, the connections between the family and the child became a site of scientific intervention. From countries as diverse as Finland, Portugal and the United States, it was believed that science would produce progress through systematic public provision, coherent policy, and rational government intervention (p. 87).

Child guidance clinics attempted to correct any dysfunctional family environment, through programs offering advice, instructions, and corrective measures targeted primarily to mothers. In the United States, women would assume leadership in campaigns to acquire better conditions for maternity and infancy through suggestions for legislative actions. An important aspect of this movement was the campaign for social benefits for mothers (e.g., mothers' pensions). Exclusionary practices were embedded in mothers' pensions discourse: mothers who were deemed alcoholics, deserted by their husbands, and single mothers did not receive benefits. To ensure monies were being spent judiciously, recipients were subjected to inspections of their homes. Mothers were penalized for using language at home other than English, for living in houses or communities considered inadequate, for improper levels of cleanliness or order at home, or even for allowing "unsavory" relatives to live in home (Rose, 1999, p. 130). At the same time, other strategies of administration were directed to the American families:

> An assemblage of connections produce a relation of the child and the family as an object of political rationality and social administration. U.S. social and medical policies, schooling, and day care and nursery schools of the early twentieth century tried to inculcate a universal, healthy child, the good scientific "professional" at-home mother, and a normal family that would be autonomous and responsible for themselves. (Popkewitz & Bloch, 2001, p. 88)

In most European countries, a central government connected a whole array of social resources for the government of poverty, insecurity, health, and education. A central issue for the social government was prevention. Child guidance clinics to promote children development, schools, industries, courts, and so on were places where specialists identified diverse problems and intervened to prevent pathologies or to correct deficiencies based on a model of a citizen responsible for himself and his family and as an active member in society. Over time, specialists produced an array of "new bodies of mundane, practical social knowledge of the habits, conducts, capacities, dreams and desires of citizens, and of their errors, deviations, inconstancies [sic] and pathologies" (Rose, 1999, p. 132).

"Problem-families" became the primary target for those specialists. The professionals had, as support, legislation and specific rules, as well as investments that permitted the use of resources such as the radio for the diffusion of expert knowledge. An important function of social governments was the concern with crucial moments of family life, such as the birth of a child, an illness, a wedding, lack of employment, and so on. As related to Brazilian education discourses, the entrance of children into school would be among those crucial moments suggested by Rose (1999). For school authorities, the first grade was an object of great concern because of high rates of failure. The entrance of children into puberty was also considered a critical moment, since it was at this age that children tended to break away from their parents' control. It was a time when youth began to prefer a colleague's company and a time when they were viewed as vulnerable and could be drawn into dangerous or illicit activities.

The Progressive School and the
Government of the "Problem-child"

How was it possible to invent and govern the "problem-child" as a distinct category from the "abnormal" child, the "dangerous" child, and the orphan or abandoned child? I understand that at least two factors contributed to this. The diffusion of schooling for the masses is the first factor. The primary school was made accessible to the whole population, an international process that took place in the late nineteenth century and early twentieth century. The second factor was the new psychological knowledge used to think about educational issues and children's development. In Brazil, these two factors were closely associated with the Progressive School movement since the 1920s. Educators defended both the democratization and expansion of the public school, and the renovation of teaching methods based on psychological knowledge about children in general and about individual differences. It was at this time, when schools were in the process of becoming more democratic and interested in meeting the individual needs and interests of children, that educators become specially committed to promoting the child's development and autonomy during which the categories of problems attributed to them multiplied. Along with the creation of new "problem-child" categories came the development of new tactics of control such as observational procedures, tests, and case studies that educators deployed to diagnose and treat problems.

The Progressive School propitiated the appearance of the "problem-child" to promote more completely the fulfillment of children's potentialities through studying, in a detailed manner, each individual. In attempting to preserve healthy development and prevent problems in the upbringing of an autonomous citizen, educators judged early intervention as necessary. A good example of this kind of concern is presented in the book *Tests ABC to verify the necessary maturity for learning how to read and write*, published for the first time in 1933.[2] In this work, Lourenço Filho, one of the most important Brazilian educators of that period, presented a test he invented to evaluate each child's degree of maturity.

In Brazil, since the beginning of the twentieth century, the first grade in elementary schools was a major concern for the school administrators. In São Paulo, for example, in the early 1930s, about 40 percent of students failed to learn how to read and write and to pass to the second grade. In Lourenço Filho's view, in order to solve this problem it was necessary to adjust teaching procedures to each child's needs. Based on his experience as an educator and in several studies developed by him and by other experts, he argued that maturity was a better criterion than chronological or mental age to evaluate a child's capacity to learn how to read and write. He concluded that there was a "general level of maturity" that was required for the acquisition of literacy and that this maturity level was not directly related to the mental or chronological age as was generally thought. Rather, maturity level was something more or less independent that needed to be measured by special means. As a result, Filho fabricated the *ABC tests*, which consisted of eight small tests to measure different aspects of "maturity," such as motor and visual coordination, vocabulary, attention, memory, resistance to fatigue, and so on.

These tests, which measured the level of maturity of each child, permitted identification of children who were not able to start learning how to read and write. For those who were not ready, he recommended prescholar activities for some months before beginning formal learning. Educators distributed the children's results on the test according to the Normal law or curve. In addition, Lourenço Filho (1933/1974) proposed that the population of first grade students be divided into three classes: the best students, the average students, and the weak students. He explained that the first group could learn to read and write easily in one semester, the second group could learn in one year, and the last group would not be able to learn in the regular period under normal conditions. For this "weak" group of pupils, educators were to provide special attention and teaching techniques.

Lourenço Filho thought that once teachers fine-tuned activities to the level of the students, schooling would be more efficient. Dividing students into three groups was also designed to prevent frustration and the development of low self-esteem in immature students who would fail to learn at the same pace as "best" or "average" students. For him, self-esteem was a fundamental element of successful learning. In this sense, it became very important not to expect of the children more than they could do. Lourenço Filho also assumed that children with low self-esteem would lose interest in learning, leading to discipline problems in the classroom.

According to Foucault, the government of populations does not limit itself to a superficial and a global level, but also acts on a deeper level and in the minutiae. As a technology created to govern first grade populations of pupils, the *ABC tests* were not limited to providing a means for dividing this population into three classes using the criterion of level of maturity. These tests also provided teachers with a means to identify the specific characteristics of their classes and major needs of each group, for example, to improve motor coordination or visual memory. With the administration of these tests it became possible to identify abilities, dispositions, and weaknesses of each individual. Teachers were encouraged to create collective or individual exercises to correct children's problems or to stimulate those who were categorized as "immature."

Moreover, during the tests, in addition to evaluating specific skills that the tests were designed to measure, the examiner was required to take careful notes about the child's behavior. In cases where teachers identified problems, they were to submit the child to additional tests and to the scrutiny of experts. If problems were thought to be related to social nonadjustment, it was necessary for examiners to produce a case study of the child that was to be "as complete as possible." This produced a developmental history of a particular child that could be used to discover the root causes for individual learning problems of that child.

From the above discussion related to the book written by Lourenço Filho (1933/1974), it is possible to discern the central aspects of the government of the "problem-child" in the early decades of the 1900s in Brazil: the idea that every "normal" child could learn if individual needs were met by the school; the combination of tests with case studies to produce objective "scientific" knowledge about children; the creation of new psychological categories to classify children as immature, passive, aggressive, with feelings of inferiority, and so on; the attempt to prevent psychological and social problems by adapting teaching to the particular conditions of the individual child and by making the learning environment as homogeneous as possible; and the efforts to help children who were considered "unready to learn."

The Government of Children Through the Family

An analysis of Brazilian education and psychology books written during the 1920s and 1930s about the problems of children allows us to observe a transitional period related to how these authors explained children's disturbances. It was during this time frame that specialists began to classify disturbances according to their causes, identifying whether they were biological or environmental, and elaborating different recommendations for prevention and treatment based on perceived causes. Thus, along with explanations based primarily on hereditary and biological factors, new explanatory discourses began to appear which valued the influence of the environment. At the same time that this shift to environmental causes was taking place, there was an increasing valorization of educational intervention as a way to correct and prevent the problems caused by an unsuitable environment.

In 1939, the book *The problem child: The mental hygiene in the elementary school*, written by Arthur Ramos, was released. Ramos, a graduate of

Medicine School of Bahia, played a large role in the development of
Educational Psychology as well as in the Child Mental Hygiene movement
in Brazil. Mental Hygiene, Ramos argued, through the study of acquired
habits of an individual at an early age, was the area of knowledge that
would ideally provide solutions for non-adjusted children through prevention and correction strategies. According to Ramos, experts had devised
the notion of the "problem child," as reflected in the title of his book, to
accurately label children who had become nonadjusted due to poor conditions in the environment where they lived.

> The concept "problem-child" was created to replace the degrading
> and strict term "abnormal child" and it was used to indicate all the
> cases of character and behaviorist non-adjustment of children to
> home, school and school curriculum. Some authors use the expression
> in a broad sense, understanding in the concept of "problem" all the
> child difficulties—physical, mental and social. The expression was
> kept, however, to name specifically the cases of psychosocial
> non-adjustments which do not come to the limit cases of mental
> disturbance. (1939, p. 21)

According to Ramos, traditional child studies overly exaggerated the
role of heredity in human development. While not denying its importance,
he asked researchers to take into account the influence of the environment,
especially the familial one, on the physical and psychological characteristics
of children. He established a direct relationship between the "problem-
child" and "problem-parents" and in tandem with many of his contemporaries, considered the family the "*fundamental social unity,*" who were
responsible for the development of the child's personality (p. 42).
The all-important concern was with poor families, as they were thought
to be unable to provide their children with a healthy developmental
environment.

Brazilian intellectuals at the end of the nineteenth century and the beginning of the twentieth century were well informed about current educational theories in other Western countries. Educators quoted books and
texts written in English, French, Spanish, and German, indicative of their
desire to participate in the international movement and incorporate into
Brazil "modern" educational thought as discussed in the United States and
European countries (Lopes, 2002, p. 320). This practice is illustrated in
Ramos' book *The problem child*. Ramos referred to Benson and Altender,
the authors of *Mental hygiene in teacher institutions*, in the United States: a
survey (1931) to affirm that "the major task of the mental hygiene in education is to preserve the normality of the normal child," emphasizing the
preventive function in relation to the corrective one (Ramos, 1939,
p. xxii). Perhaps the origin of the term "problem-child" in his book is related
to the appearance of child-guidance clinics in the United States as Ramos
makes reference to a 1936 book, *Problem child*, by John Edward Bentley,

which discusses the functioning of these clinics in the United States. As Patto observes,

> His book [*The problem child*, by Ramos] is part of an extensive international literature that, in the 1930s and 1940s, brought in the title the expression "problem-child," had as key-word the concept of maladjustment and as objective the correction of the deviations. (1990, p. 192)

It was during this era that psychoanalysis was introduced in Brazil and, according to Lopes, arrived as part of an alliance between psychiatry and the state with the goal of elaborating a preventive project of public hygiene in the urban centers. These effects were primarily directed at the Imperial period's orphans, recently freed slaves, Indians, poor whites and immigrants (2002, p. 321). Ramos contributed to the dissemination of psychoanalysis in the educational field through the publication of books such as *Education and Psychoanalysis* (1934) and *The problem child* (1939). On the one hand, along with his contemporaries, Ramos argued that education should meet the pupil's individual characteristics. On the other hand, and in conjunction with educational thought widespread during this time, Ramos highlighted that the most important objective of this investment was not the individual, but society. In his book *Education and psychoanalysis*, he stated: "Being directed to the individual, education looks forward, however, to the society. And it's last effort will be in obtaining greater social results" (p. 14). Educators saw the prevention of children's problems, therefore, not only as a measure that sought to solve individual difficulties but, simultaneously, as a way to strengthen society.

Drawing on Adler, Ramos affirmed that the schools' primary function was to correct, in the child, the excesses of the "will to power" and develop a "sense of community." Children who were thought to present major problems were those who had "inferiority of organs" or were spoiled or, to the contrary, were hated by their families or peers—as they did not have a sufficiently developed "sense of community." For Ramos, the significance of individual psychology was its contribution to correcting familiar and school problems (1934, pp. 54–56). To develop a "sense of community," then, was one of many theoretical tools that allowed Ramos to make a logical connection between individual expectations and social requirements that would justify intervening in the family to save the child.

In his book, *The problem child*, Ramos discussed several examples of learning problems studied in this institution. According to Dante Moreira Leite, Ramos's book had been, for some years the only book available in Brazil that presented an empirical study on learning problems (Patto, 1990, p. 80). In the section of this book dedicated to the possible causes of learning problems children displayed in school, Ramos sought to show that difficulties originated mainly from the "problem-child's" family dynamics. Based on the theories provided by psychoanalysis and Adlerian psychology, and alluding to American, French, German, and other foreign authors,

Ramos argued that the quality of care that a child received at home, specially from the mother, was a determining factor in shaping that child's future adjustment to the school and to the broader social environment. In *Education and psychoanalysis*, Ramos posited that

> The mother should be naturally the first educator, with the father's help; she should appear to her son as the first other, and then awake in him the interest in the others: father, siblings and people in the familiar environment, at first, and in the social environment, next. (1934, p. 58)

When studying the child's problems, Ramos (1934) did not give preference to any special method or technique, but relied on a variety of techniques—"incidental observations, biographic fragments, systematic observation, questionnaires, case's histories, tests and measures, experimentation, etc" (p. 23). The examples presented in the book reveal that professionals collected information such as data about the child's parents, brothers and sisters, the child's birth's circumstances, diseases, sleeping habits, and housing conditions. They also searched for information in the school and, based on the data collected, made recommendations to the family and to the teacher, asking for changes in the child's habits and in the ways in which the adults related to the child. To avoid sexual problems, for example, Ramos presented a list of recommendations that related to the adequate organization of the familiar intimacy, particularly the relations between the child and the mother.

> Avoid continued spoiling and stroking, do not breast feed beyond the normal period, avoid the use of pacifier, separate as soon as possible the child of the parents' bedroom, do not allow that the child sleep in the same bed, avoid the marital intimacies in the child's presence, adopt a natural aptitude in face of the manifestations of sexual character, presented by the child. (Ramos, 1939, p. 315)

Advice such as the above was given to parents at the Mental Hygiene Service as a way of intervening in the families' lives in order to correct bad habits that could lead to child's maladjustment to the school.[3] The rule that the child should sleep in his/her own bed and not in the parent's bed is an example of a ubiquitous recommendation, made by Ramos based on psychoanalytical knowledge of Freud and Adler and illustrates the practice of biopolitics in regulating the lives of pupils, including the home life of parents.[4]

In the conclusion Ramos (1939) defended the idea that the "problem-child" whose difficulties originated in a maladjusted family environment should attend regular schools. For it was in the schools that these children could be assisted by hygienists and other professionals such as doctors, psychologists, and social workers who would contribute in their recuperation. For Ramos, the first set of causes to be identified and treated by the experts in the Mental Hygiene Clinics were those classified as "medical–organic" problems. If the physical problems were not identified and treated, it would be difficult to treat any mental hygiene problems. It was necessary to start, then, with the most apparent problems, those that were manifested in the child's body. Following this, it would be possible to reach the subtler plane of her/his emotions and soul. Importantly, it would also be necessary to intervene in the family and provide parents with explanations about the causes for their children's problems as well as advice on how to correct them. However, this was considered an extremely difficult task, since parents were unwilling to recognize they were responsible for their children's difficulties at school. Finally, at times it would be necessary to intervene at the level of the school. For example, in cases in which a child was hated at home, Ramos stated that the teacher should understand the child and act as a substitute for the mother. The teacher was to assume a pastoral function—taking care of the soul and the salvation of each and all of her pupils. Once again, the theoretical basis for the exercise of this power was psychoanalytical knowledge.

It is not possible in the scope of this chapter to discuss each aspect of the government of the child as it was configured in the services described by Ramos in *The problem child* (1939). But from my discussion above, it is possible to observe that problems were found in the everyday life of the everyday child, marking an important difference between the "abnormal" child and "problem-child." I argue that the use of the expression "problem-child," in Brazilian educational discourses suggested an expansion of the idea of "nonnormality," that eventually transformed every child into an individual who, under bad circumstances, could develop problems of adjustment. It was not only the poor, the abandoned, the ill, or the defective child that experts included—although these populations were still the main objects of concern and subject to more strict interventions—but also the "well-raised" child who was now subject to scrutiny.

The figure of the "problem child" allows us to identify a significant transformation in specialists' explanations of children's difficulties and their recommendations for dealing with children's schooling problems. Instead of considering them to be biological defects, the problems would be framed within the emotional and the social domain. Because of this transformation, treatment would be thought about mainly in terms of therapy directed to the family. Instead of the exclusion of the defective child, it would now be possible to think of educating families to create the suitable homes where the child could develop and became well adjusted. As I have already mentioned, hygienists defended the idea that the most important task of mental hygiene was to preserve the normality of the normal child, which meant to preserve the biologically healthy child, avoiding the appearance of vices of conduct and improving the conditions for an adequate adjustment to the social environment initiated in the school. Educational experts believed that many of the most common types of maladjustment were due to a bad constitution of the family environment and stated that it was this that should be cured by means of the transformation of the familiar habits, especially those related to the interactions between parents and children. "Solved the causes, ceased the effects" was a ubiquitous sentence in the educational literature. In this enterprise, all the factors that influenced the child's life conditions should be observed. The well being of the child was linked to the optimum functioning of society, such that whatever was proposed to help the "problem-child" would, at the same time, improve social organization.

Concluding Remarks

In this chapter, I sought to indicate the means through which the discourses produced by, and available to, educators since the 1930s made the "problem-child" visible and governable. It was possible to verify how the child was "mapped" by practices of observation of his/her attitudes in the school, the deployment of psychological tests, and the elaboration of case studies. From this "data" multiple causes of the school maladjustments were discerned. Categories escalated, as deeper and subtler aspects of the personality were fabricated. The techniques recommended for governance of the "problem-children" included the identification of causes and attempts to solve them, according to the expectation that if the causes were eliminated, the effects would vanish. To prevent maladjustments, it was necessary to intervene in families whose inadequate organization was considered to be one of the main causes for the difficulties that children presented in school. It was also necessary, both at school and in the home, to offer good examples and to stimulate practices of confession and self-examination, as ways to lead the children in distinguishing the difference between right and wrong and to consciously and freely choose honesty, solidarity, justice, healthy habits, and so on.

Multiple knowledges illuminated and justified pedagogical orientations about suitable ways to deal with maladjusted children. Knowledge from biology and hygiene, moral principles, psychological and psychoanalytical "discoveries" about children's "functioning" combined in various ways to produce explanations and recommendations about how to treat different types of deviations. Children were conceptualized as incomplete and developing beings who are more permeable to environmental influences than adults. In this sense, experts considered it necessary to provide children with physical and social spaces favorable to healthy biological, psychic, and social development. The type of citizen that educators had in mind was autonomous and sociable. It was in planning the preparation of this ideal citizen that specialists should collaborate with the family and the school. Modifying the conduct of the "problem-child" required, first, a transformation of both the conduct of the mother and teacher. Following the introduction of psychoanalysis, that meant not only modification of how adults dealt in interactions with the child, but also changes in how adults dealt with themselves, with their histories, with the memories they had of their own childhoods, of their parents and so on. In this sense, governance of the child through the family entailed both governance of the family and also governance of the teacher through the child.

In March 2005, the famous Brazilian educational journal *New School* released the article "How to create a cozy school," discussing how teachers should deal with emotions in the classroom. Teachers are advised to use strategies such as proximity, listening, valorizing children's capabilities, believing in the child's potential to improve the child's self-esteem in order to promote the formation of happy and responsible citizens. In this article, problems of aggressive behavior and difficulties in making friends are explained as consequences of inadequate maternal attention and other family difficulties experienced by the child. The following depiction of a "problem-child" is a typical example: "Son of divorced parents, he lives with the mother—who works too much and does not have time to look after him. Without vigilance, the boy spent most of his time in the streets and was almost always absent in school" (Cavalcante, 2005, p. 55). The school's principal describes the initial behavior of the student as aggressive and added that this could be explained by his life history.

The language in the above article is similar to the language published in the 1930s and 1940s on "problem-children." Today's "problem-child" is still explained by reference to psychological problems located in the child and related to a maladjusted familiar environment. As in previous decades, educational specialists continue to state that teachers should get close to the pupils and encourage them to verbalize their feelings and worries. Recommendations dealing with pupils' lies and robberies are also similar: "If a pupil lies, for example, the ideal is to talk privately with him, asking for the consequences of this aptitude, for the way his colleagues would act when they discover the truth" (Cavalcante, 2005, p. 55).

On the other hand, there are some differences between the specialists' discourses from the 1930s and 1940s and those of the present. Although for many educators the family remains the origin of students' problems, several educators have contested explanations that locate the problems in the child. The Brazilian educator Sandra Maria Sawaya (2002), for example, states that the educational system itself produces many problems by overloading teachers with bureaucratic tasks, creating obstacles to collaborative work, and imposing strict disciplinary practices on teachers and pupils. She proposes a kind of psychological intervention in schools that creates "speaking spaces" where various actors, the principal, coordinators, teachers, pupils, parents, and the other people that work in the school, can express feelings, discuss experiences, practices and beliefs, as well as propose alternative ways of dealing with problems and conflicts. Sawaya argues that an important contribution of psychology for the field of education would be to listen to the teachers about their practices, their conceptions about their work, their pupils and their relations in the school, as a way to increase their self-knowledge and their knowledge about the children's capacities and needs.

According to Sawaya, as well as other contemporary Brazilian authors, the main question is no longer to understand behavioral problems and maladjustments in school as a consequence of the pupil's inadequate familiar dynamics. Instead, problems are now seen as a product of the relationships between subjects in the context of the school. Difficulties are understood as conflicts that can be solved with the psychologist's intervention. Although from this perspective not every child is seen as a potential "problem-child," everyone should still be under *psy* supervision since conflicts can emerge from relationships and, if not solved, may cause behavioral problems in the school. In this sense it is possible to state that the *psy* gaze is now even more widespread. It is not only the children who could become involved in conflicts but all school personnel including teachers. To prevent the emergence of a "problem-school," then, it is necessary to make people talk about their feelings. And they may need the help of a psychologist to do so.

Notes

1. In Brazil, Progressive School thought was imported from United States and European countries. One of the most important representatives of this movement was the educator Anisio Teixeira, who went to United States in 1927 to study with Dewey. The renovators believed that pedagogy should be based in scientific knowledge about children's psychology. They supported the idea that it was necessary to make the child the center of the pedagogical process and considered that teaching methods needed to meet children's interests and give pupils the opportunity to learn by doing, among other principles.
2. Quotations presented in this text are from the book's 12th edition (1974).
3. Arthur Ramos was the director of the Orthophrenics and Mental Hygiene Service, created in 1934 in the Brazilian Federal District.
4. Other problems identified are "the turbulent child," the child who had twitches or other small vices—for example, one who bit his/her nails—"the drop-out child," or those with sexual problems, fear and anxiety, and children's pre-delinquency: lie and steal (From: Table of Contents).

References

Cavalcante, M. (2005). How to create a cozy school. In *New School 20* (180), 52–57.

Dean, M. (1999). *Governmentality: Power and rule in modern society*. London: Sage Publications.

Foucault, M. (1979). *The history of sexuality: An introduction*, Vol.1 (R. Hurley, Trans.). London: Allen Lane.

Foucault, M. (1983). The subject and power. In H. Dreyfuss, & P. Rabinow (Eds.), *Michel Foucault: Beyond structuralism and hermeneutics* (2nd. ed.). (With an Afterword by and an Interview with Michel Foucault). Chicago, IL: The University of Chicago Press.

Foucault, M. (1988). Technologies of the self. In L. H. Martin, H. Gutman, & P.H. Hutton (Eds.), *Technologies of the self* (pp. 16–49). London: Tavistock.

Lopes, E. M. T. (2002). The psychoanalysis applied to children in Brazil: Arthur Ramos and the "problem-child." In M. C. Freitas, & M. Khulmann Jr. (Eds.), *The intellectuals in the history of childhood*. São Paulo: Cortez.

Lourenço Filho, M. B. (1974). *Tests ABC: To verify the necessary maturity for learning how to read and write* (12th ed). São Paulo: Melhoramentos. (Original work published 1933)

Patto, M. H. S. (1990). *The production of the school failure: Stories of submission and rebelliousness*. São Paulo: T. A. Queiroz.

Popkewitz, T., & Bloch, M. (2001). Administering freedom: A history of the present—Rescuing the parent to rescue the child for society. In K. Hultqvist & G. Dahlberg, (Eds.), *Governing the child in the new millennium* (pp. 85–118). New York: RoutledgeFalmer.

Ramos, A. (1934). *Education and psychoanalysis*. São Paulo: Companhia Editora Nacional.

Ramos, A. (1939). *The problem child: Mental hygiene in the primary school*. São Paulo: Companhia Editora Nacional.

Rose, N. (1999). *Powers of freedom: Reframing political thought*. Cambridge: Cambridge University Press.

Sawaya, S. (2002). New perspectives on the success and the failure at school. In M. K. Oliveira, D. T. R. Souza, & C. Rego (Eds.), *Psychology, education and the contemporarie's life thematics* (pp. 197–213) São Paulo: Moderna.

PART 4

Governing the Modern and Post-Modern Citizen and Nation through Universal Reforms in Education

CHAPTER 9

No Child Left Behind? The Specters of Almsgiving and Atonement: A Short Genealogy of the Saving Grace of U.S. Education

KAREN S. PENA

The subject of this chapter is the construction of the child, but not in all its medical, juridical, and psychological fullness.[1] It does not seek to identify all the issues of identity, knowledge, and power but instead analyzes the space where the new reform education intersects the construction of the child as salvation, as the saving of souls. The U.S. legislation, No Child Left Behind Act of 2001 (NCLB), its updates, and the discourses associated with its passage and implementation are the practices I examine for sketching out the network of relations, techniques, and technologies which construct children for the purpose of saving them. The strategies of reform aim for a homologous construction of children, molding them as the same, by restructuring heterogeneous[2] elements selected from multiple discourses of salvation. By analyzing the language of NCLB as reconstructing the education of children as salvation, I interrogate how strands of religious discourse, which historically have referred to practices required to save one's soul for eternity, are recycled in today's secular educational discourse to save the child in this life. The title of the legislation itself recalls the enlightened humanistic aspiration that all children shall be saved to lead a useful life. At the same time, I analyze how the practices of NCLB resonate with other historical concepts of salvation that recall older formulations of salvation, which require one to prove oneself free of sin or face the possibility of failure, separation, and damnation.

Situated in the writing of Michel Foucault, and related works, I interrogate the effects of NCLB on the restructuring and governance of childhood. The idea that particular constructions of childhood are forever true is

brought into question. Following Foucault, I understand meaning[3] to arise
in the spaces between existence and the stories we tell about it. There is no
prior to the language of debates that constructs its objects of discourse;
childhood is a moving object of discourse that emerges from the educational debates that are ongoing. There is no universal truth to which the
meaning of childhood is directed, nor is there an end to which it aspires
(Foucault, 1972). There are only narratives with the practices attending
them that are retold and folded into what we take as the materiality of the
child. It is in mapping the statements and the cultural practices that we can
trace the imaginations from which they emerge. In supporting this view,
Foucault said, we "must articulate a philosophy of the phantasm that cannot be reduced to a primordial fact [of existence] . . . , but rises between
surfaces, where it assumes meaning, and in the reversal that causes every
interior to pass to the outside and every exterior to the inside" (1977,
p. 169). It is impossible to describe childhood as existing before thought or
in its essence. Rather, childhood is a cultural event in a specific time/place
and this is what produces local knowledge of the child. The child emerges
differently within different rationalities, rules of formation, conditions of
possibility, and confluence of forces (Foucault, 1980).

I first comment on the history of the four elements of NCLB:
accountability, choice, flexibility, reading, as well as the changing constructions of children suggested by the legislation. I then analyze the restructuring of the patterns of power and knowledge that shape the child in current
discourse. Finally, I discuss the ways in which shifts in the secular meanings
of "salvation" are circulating in the language of current educational reform.
I analyze the patterns in the prescriptions of NCLB as reconfigured, secularized forms of salvation that were formulated in earlier religious discourses
which formulated different religious concepts of salvation.

I argue that there are many conflicting meanings and practices of
achieving "salvation" and none of them should evoke warm and comfortable feelings of faith. The technologies of salvation are disparate and, at
times, mutually exclusive. All are dangerous. The conceptualizations we
create of children are the sites of the workings of power and knowledge
which require a shift from thinking of power as repressive to thinking of
the relations of power and knowledge, in statements and practices, as being
productive of new ways of thinking about childhood. Additionally, I
explore how technologies are not new, but instead are reconfigurations of
strategies that have historically been used again and again. They are not
readily visible in current reform because they are often disguised in the
sheep's clothing of the reform discourse itself.

Commentary on the Legislation of *No Child Left Behind*

NCLB is self-proclaimed to "promote educational excellence" for an
estimated 46.5 million U.S. public school children in 89,599 public

schools. It addresses the salvation of children in two ways. First, it defines children collectively as a population at a distance in terms of improved school performance and secondarily, individually, as a child not trapped in a failing school. In the Executive Summary of NCLB (U.S. Department of Education, 2002), the legislation "reflects a remarkable consensus . . . on how to improve the performance of America's elementary and secondary schools while at the same time ensuring that no child is trapped in a failing school" (¶2) . There are four areas of reformed practice outlined in the legislation: accountability, choice, flexibility, and reading. I will review some historically received meanings of each of these elements followed by the current shifting meanings as they are developed in NCLB. By so doing, I illustrate that the historically received meanings are radically transformed in NCLB.

Accountability

The first element is accountability. Traditionally, accountability has been a qualitative measure of responsible moral action, associated with praise or blame. Accountability was a mark of virtue associated with esteemed conduct. As John Locke (1690/1961), the seventeenth century English philosopher wrote,

> Virtue is everywhere that which is thought praiseworthy; and nothing else but that which has the allowance of public esteem is called virtue. Virtue and praise are so united that they are called often by the same name . . . it is no wonder that esteem and discredit, virtue and vice, should be a great measure everywhere correspond[ing] with the unchangeable rule of right and wrong (pp. 176–177).

Accountability was a measure of the quality of one's conduct, one of several moral dimensions of one's being that was in public view.

In NCLB, accountability is transformed to a quantitative measure. It is literally achieving the standard score in reading and math tests. Each child is accountable, but so are the teacher, the principal, the school, and the state. The child's score brings salvation or damnation not only to himself, but to all levels of the educational bureaucracy. Accountability bleeds through the cracks of bureaucratic structures so that it is everywhere. Each state, on the basis of federal law, is responsible for implementing the testing of each student in reading and math and to measure each student's performance against a "legitimate" state standard. This is the measure and the emergence of the child as a score. Between the federal government and the child are the state, the school district, the school, the principal, the teacher, and the parents. On the one hand, it is the child who must do the work of meeting these standards. On the other hand, the effects of each child's score rain down on all the players, not as individual scores, but as populational divisions, such as socioeconomic, racial, language and disability categories. Accountability is at once the responsibility of the child but not only the individual child. According to the Executive Summary:

> The NCLB Act will strengthen Title I accountability by requiring States to implement statewide accountability systems covering all public schools and students. These systems must be based on challenging State standards in reading and mathematics annual testing for all students in grades 3–8, and annual statewide progress objectives ensuring that all groups of students reach proficiency within 12 years. Assessment results and State progress objectives must be broken out by poverty, race, disability and limited English proficiency to ensure that no group is left behind. (U.S. Department of Education 2002, ¶4)

Redemption is the status of being saved from failure. It is achieved when collective test scores in reading and math do not fall below state proficiency norms or standards. Although the title of NCLB speaks only of the child, the subjects of the law are as much the state, school districts, and individual schools. According to the Executive Summary, "school districts and schools that fail to make adequate yearly progress (AYP) toward statewide proficiency goals will, over time, be subject to improvement, corrective action, and restructuring measures aimed at getting them back on course to meet State standards." Children appear as test scores in the aggregate which can bring the failure or the awards of achievement to schools. Therefore, "Schools that meet or exceed AYP objectives or close achievement gaps will be eligible for State Academic Achievements Awards" (¶4). Children, in the NCLB discourse of accountability, are constructed as flat and faceless objects who achieve salvation by meeting standards of proficiency on reading and math tests. Children are to construct themselves according to these standards of conduct, which are an unexamined tension between children as a population (group test scores) and children as individuals, therefore as individual test scores.

Choice

The second element of NCLB is choice. Historically, choice has been envisioned as the course of action one takes on the basis of the knowledge one has of the world. It was characterized, as Locke (1690/1961) observed, by "deliberation and scrutiny" (p. 121). Choice appeared in humanistic thinking, and was taken-up as the measure of an individual. Locke found that "Thus the measure of what is everywhere called and esteemed virtue and vice is this approbation of dislike, praise or blame, which by a secret and tacit consent establishes itself in the several societies." (pp. 175–176). Particular acts may be judged differently, yet choice among possible actions was a moral responsibility requiring knowledge of the world. Choice has also represented the creation of diversity and difference, which have expanded opportunities and variety in patterns of thought, lifestyles, and beliefs. In education, choice has represented the decisions children might make in their eventual selection of courses of study, professions, or vocations. Pedagogically, there was choice in the selection of various curricula, aims, objectives, and alternate styles of instruction. Historically, choice has been the various forms of knowledge and technologies of expanding opportunities and increasing diversity along many trajectories.

In NCLB, choice is transformed into what can be exercised by a student if a school fails to meet state numerical standards of performance. Choice is constructed in the language of lack, deficiency, and deprivation. For choice to appear, there must be constructed the distinction of diversity, not as abundance, but as one who is less than capable (Dussel, 2002). There must be failure to exercise choice; this means that difference is recognized only as falling below standard. Difference is therefore bad. Students and schools that are identified as failing are enclosed into a category of "in need of correction." NCLB purports that choice is a path to a restoration of a level of student performance from which one has fallen. Choice offers a "relief" for a "trapped" child, who, without that choice has lost his chance for redemption. It is a process of "justification" or a return to a status of saved. The new conception of this normal status for children is a "universalized" conception of making students the same. The acceptable measure of student capability in relation to standards is what constitutes salvation. The exercise of choice is an effect of the responsibility of the state implied in the language of the legislation. Accountability and responsibility are joined: there is a conflation of the child, the school, and the state. Implied in the language is also the supposition that once choice is exercised, the student will continually work upon the self in order to develop his natural competencies and potentialities. In other words, the student is expected to behave responsibly in his own self-actualization. NCLB advocates that the responsible action to take, both for the state and for the student or his parents, is to restore each child to a natural path of achievement. NCLB authorizes students attending a persistently failing school to receive supplemental educational services and transportation from local educational agencies. Choice is an instrument envisioned as providing an opportunity "to help participating students meet challenging State academic standards" (U.S. Department of Education 2002, ¶7). Choice is a technology or practice that carries within it the assumption that it constitutes, at least in part, the path to salvation.

Flexibility

The third element of NCLB is flexibility. Flexibility has been conceptualized historically in various ways. It has meant the variety of possible educational aims, objectives, and instructional activities adapted to the interests and needs of children. For example, Locke (1693/1996) wrote in the tradition of humanistic education:

> He, therefore, that is about children should well study their natures and aptitudes and see, by often trials, what turn they easily take and what becomes them, observe what their native stock is, how it may be improved, and what it is fit for; he should consider what they want, whether they be capable of having it wrought into them by industry and incorporated there by practice, and whether it be worthwhile to endeavor it (p. 41).

Locke writes at length about how a teacher needs to design and shape both curricula and instructional practices to the diverse needs, abilities, and interests of the child. Flexibility for Locke resides in adjusting educational practices to how much the child wants to work on something and to the how efficient the time spent might be. It might be better to work on something else at a given moment.

In addition to pedagogical flexibility, flexibility, in the past, has also referred to a sociality of shared responsibility where individuals work collectively in teams or partnerships to solve problems (U.S. Department of Labor, 1991). Flexibility has been understood, more recently, as the construction of a self that can adapt to the demands of a changing workplace, learning new skills, and shifting vocations as the world is reconfigured. Finally, flexibility has referred to the use of self-knowledge in the construction of a reflexive self who assumes the responsibility of shaping his individuality and personality into a self-disciplining, desirable lifestyle (Foucault, 1988; Hunter, 1992). As Ian Hacking (1995) noted, life is lived within the possibilities of this life: "These ancient values are ideals that none fully achieve, and yet they are modest, not seeking to find a meaning in life beyond life, but finding excellence in living and honoring life and its potentialities" (p. 265). This kind of thinking takes a standoff external view of the self in which the self analyzes the self in a reflexive analysis. Hacking defines this kind of flexibility as a way of founding the construction of one's being.

> Self-knowledge is a virtue in its own right. We value the way in which people can fulfill their own natures by gaining an unsentimental self understanding. We think it is good to grow, for all our vices, into someone who is mature enough to face the past and the present, someone who understands how character, in its weaknesses as well as its strengths, is made of interlocking tendencies and gifts that have grown in the course of a life (p. 265).

Hacking's view of the flexible self is not the kind to be found in NCLB. There is today a transformation in the kind of flexibility that emerges with NCLB that picks up on a theme from behaviorism. There had been, since the beginning of the twentieth century, a forfeiture of conscious experience in the construct of the self as, for example, explored by Benjamin Franklin in his eighteenth century *Autobiography* (1961), in favor of behavioral performance (O'Donnell, 1985). This is a move that brought populational reasoning to dominance in educational reasoning in the problematization of the child. Populational reasoning is the understanding of the child's performance through norms and statistical analysis. It is populational reasoning that has become incorporated into pedagogical thought about the child. NCLB utilizes this perspective of children, and transforms the construction of children even further to a space where reading and math skills are no longer something constructed outside the individual and selected for instruction. Reading and math performance standards are assumed in the legislation to be natural forms of behavior in the developmental process; they are constructed as already internal and part of the unfolding nature of the child. Children are in this sense the "same" equivalent with one another; difference becomes a difficulty. Popkewitz (1998) explained this new form of flexible individual where external norms of the "responsible subject" are made to seem internal and operate at "multiple layers" of self-administration. No longer are children correlated with abstract norms:

> Instead, populational reasoning individualizes the norms as though the norms exist internal to the child. This type of individualization is an architecture of regulation as it no longer seems socially constructed but a systematic "knowledge" about the norms in which the child develops "self-knowledge" (p. 132).

In NCLB, flexibility, no longer possible as diversity, is radically transformed into a series of financial options in "exchange for strong accountability for results." States or local educational agencies can form partnerships to restructure the use of state funds within the frame of existing programs. Up to 50 percent of funds from other federally funded programs may be transferred from one program to another. The consolidation of funds at both the state and local level require "performance agreements with the Secretary of Education." NCLB joins salvation of the child to statistical performance standards in reading and math and directs its resources to that end.

Reading

The fourth element of NCLB is reading. The legislation intends that every child shall read by the end of the third grade, and this unitary, singular goal will be assured by the use of scientifically based reading instruction programs. The assumption in NCLB is that science already knows how to bring the reading performance of the child to an acceptable standard. NCLB understands reading as a linear, unitary "thing" which will appear through the proper application of scientific technologies. This assumption forgets that reading was constructed by human beings. It forgets the multiple readings of reading, as Derrida (1967/1976) explores. It forgets the heterogeneity of skills and complexities of the formation of meanings. By directing all children to the same particular standard based on grade level, NCLB forgets the diversity of the histories and individual life trajectories of students. It forgets the idea of diversity of talents and abilities, which has circulated from antiquity. Diversity disappears when attention is focused on the same.

NCLB considers reading skills part of human nature. This is not the first time knowledge was understood to be natural and universal content necessary for the education of children. In other times, particular knowledges were self-proclaimed to have an inside track on the natural meaning of life for children, but reading was never more than a tool for unlocking truth. One example of the internalization of the source of knowledge is found in the thinking of traditional Calvinism, where an individual child explored the meaning of man's covenant with God. One way of discovering the depths of immortal truth was through reading Scripture. Another example of capitalizing on what was believed to be the natural inclinations of the child was the thinking of G. Stanley Hall (1904/1908). For him the child already contained the history of the race; education raised them towards higher realms of development. Hall said that the child must:

> enter upon a long viaticum of ascent, must conquer a higher kingdom of man for himself, break out a new sphere, and evolve a more modern story to his psycho-physical nature. Because his environment is to be far more complex, the combinations are less stable, the ascent less easy and secure; there is more danger that the youth in his upward progress, under the influence of this 'excelsior' motive, will backslide in one or several of the many ways possible. New dangers threaten on all sides. It is the most critical stage of life, because failure to mount almost always means retrogression, degeneracy, or fall (pp. 71–72).

For Hall, reading was a time of sampling and suggestion not subject to examination.

> There is now evolved a penumbral region in the soul more or less beyond the reach of all school methods, a world of glimpses and hints . . . It is the age of skipping and sampling . . . What is acquired is not examinable but only suggestive . . . That is why examinations in English, if not impossible, are very liable to be harmful, and recitations and critical notes an impertinence, and always in danger of causing arrest of this exquisite romantic function in which literature comes in the closest relation to life, keeping the heart warm, reinforcing all its good motives, performing choices, and universalizing its sympathies (pp. 474–475).

For Hall, reading was constructed as instrumental in the unfolding of a veiled human nature. It served to influence the development of the mysterious, hidden recesses of the soul.

Two other examples of educational perspectives whose constructions of reading and the child claimed an unfolding of human nature are, first, expressionism noted by Rugg and Shumaker in 1928 as:

> The creative impulse is within the child himself. No educational discovery of our generation has had such far-reaching implications. It has a twofold significance: first, that every child is born with the power to create; second, that the task of the school is to surround the child with an environment which will draw out this creative power. (Cited in Cremin, 1961, p. 207)

This would "take the lid off" the child and through the release of their creative impulses remove the fear of the future.

The second example is the Freudians who took the unconscious, repressed emotions, transference, and sublimating seriously. For example in 1928 Margaret Naumberg at the Walden School, provided a curriculum built on the

> apparently unlimited desire and interest of children to know and to do and to be . . . For to us all prohibitions that lead to nerve strain and repression of normal energy are contrary to the most recent findings of biology, psychology, and education. We have got to discover ways of redirecting and harnessing this vital force of childhood in constructive and creative work . . . [abandoning a] false dependence on the glib authority of teacher or textbook . . . [and seeking instead to nurture in children an indomitable] independence of feeling, thought, and action. (Cited in Cremin, 1961, pp. 211–212)

Often in the Walden School, where there was no written curriculum, resources and activities were drawn from what students wanted to do. Reading was one of many possible activities. In these examples, reading is a flexible technology for learning other things; it is not an end in itself.

In NCLB, reading becomes a static, graded concept. This notion can only appear and function if children of different age groups are thought of as possessing equal capability. The new reform does remember an old sacred understanding from the reformation: the equality of souls. But this notion has to be reconfigured as equal ability. The equality of souls is transformed in populational reasoning as the return of the Medieval idea of equivalent capacity (Aries, 1962/1964, p. 187). Diversity as normal, natural and indeed desirable has to disappear in order for children to be recognized as being of "equal ability." The multiple forces historically constructing diversity in children are collapsed into a concept of children as the same.

Being for the child is transformed in NCLB even further by relations of power and knowledge into the same, the same standards of reading and math performance.

Salvation as Patterns of Governance and

Relations of Power/Knowledge

Power is organized and circulates in networks of relationships. As power is exercised through apparatuses of bureaucratic controls and subtle mechanisms of gesture and desires, it evolves, organizes and puts into circulation, particular knowledges that are taken as true. It is within the circulation of forces of competing discourses that the individual emerges as a subject. There is a certain economy of discourses; otherwise it would not be possible to have a convergence of forces, or a field of forces, for the exercise of power/knowledge in the production of discourses of truth. From this perspective, the child can be understood as being simultaneously a construct of power/knowledge circulating in competing discourses and at the same time a vehicle for continuing debates for normalizing truth about the child (Foucault, 1980).

NCLB is a site of discourses circulating debates of truth of the child. It continues in the general pattern of linking educational discourses with those of political authorities, administration, social sciences, competence and the construction of subjectivity. It does not function as a separate set of practices, but resonates with other discourses that construct the child and are constructed by the child. For example, NCLB correlates with the psychological and administrative discourses of division and distinction which interpret educational practices as rescue (Baker, 1998). It is homologous with discourses in other institutional fields that emphasize accountability, choice, flexibility, partnerships, and responsibility, such as business and the military. It continues the discourse of enclosing ethics into the practices of science as freedom. Following the thinking of Nikolas Rose (1990), the language of NCLB locates ethics in education as science, resulting in "tying the norms of truth to education and youth, it binds subjects to a subjection that is the more profound because it appears to emanate from our autonomous quest for ourselves, it appears as a matter of our freedom" (p. 256).

Increasingly, the forces that constitute the field of formation for current constructions of the child intersect in the body. The ideas of salvation of the child are no longer about spiritual redemption which assures the soul of a pleasing existence in life after death throughout eternity. The constructions of salvation have shifted to concentrating the forces of reform in the body of the child to insure a self-governing, productive existence in a secular, globalized world.

NCLB constructs the child as operating in a field of forces, partly determined by habitus, which limits what can be thought and done and conversely what cannot be thought or done. The interplay of global demands and local control situate the child as experiencing education in the body. Pierre Bourdieu (1997/2000) explored these relations of power and the construction of self-knowledge. Exposed to the world as "sensation, feeling, suffering, it is the engaged body that takes hold of the world and masters it. It is the body as an instrument which behaves as if things were transparent; it performs as a seemingly autonomous agent" (pp. 142–143). The power and knowledge that shape and are shaped by a child are taken for granted, "precisely because he is caught up in it, bound up with it; he inhabits it like garment or a familiar habitat. He feels at home in the world because the world is also in him" (p. 143).

This embedding of the construction of the self in cultural knowledge is a way of understanding the way children are caught up in the language of NCLB. The new reform lies in a space opened in the transforming relations of power and knowledge. The transformation lies in the gap, experienced as a positive or negative surprise between expectations and experience. NCLB identifies this gap as the difference in the bargain "first struck by President George H. W. Bush" in 1989 that authorized "waivers" and the current reform of NCLB, which exchanges educational funds for "strong accountability" (¶10). The new penalties and punishments in NCLB, for example the closing of schools and the recycling of students, are meant to increase the obligations of children and enforce their obedience to the rules of governing. In their implementation, they are conflated and reduce the child to simple proficiency in reading and math.

NCLB draws the whole family into its schemes of accountability. The divisions and distinctions that the legislation creates and then punishes restructure our understanding of the child and the family. Jacques Donzelot (1977/1997) described a double-edged shift in family life which occurs with new exterior norms; first, there is the withdrawal of the family into itself with the reduction of its autonomy, and second, there is an intensification of family life. According to Donzelot, subjectivity in the advanced liberal family is torn by the "centrifugal and centripetal forces" in which it is caught:

a sort of endless whirl in which the standard of living, educational behavior and the concern with sexual and emotional balance lead one another around in an upward search that concentrates the family a little more on itself with every turn; an unstable compound that is threatened at any moment with defection by its members (p. 228).

The increased effects of the power of the state in constructing knowledge through state standards are felt in the construction of subjectivity as a body within the body of the family. In Donzelot's thinking, "by targeting the body, power risks separating the living from what makes one want to live" (p. 234). There is, in Donzelot's thinking, a loss of family unity and tradition as the family turns to the wider demands of global/local relations of power and knowledge.

Nancy Lesko (2001) echoes this sense of loss or abandonment. But instead of a sense of loss in the reformulation of the child/family relations through shifting power/knowledge, Lesko addresses a sense of loss in the reformulation of the meanings of the child, the reformation of the stages of child development, and the effects on the body of the developing child. In her view of the new relations of power and knowledge, the child is being fast-tracked to adulthood. The construction of the child has moved from being "cared for" to the child becoming a productive individual, who is evolving to a citizen, responsible for governing himself and living as an independent being in the world. The extended years of the next stage of development, adolescence, are being squeezed, by requiring children to more quickly become adults. The less than capable performers, the deviant and difficult, are enclosed as the dangerous and encumbered with a series of legislative penalties:

> As children below ten years of age have become erotic, spectacular, and marketable, the teenager's market share has sunk. Slow development in time may no longer be functional, and quick leaps from childhood to adulthood may be called for by virtual workplaces and education provided on line (p. 198).

Lesko notes there was a turn in 1996 from a time when children were thought of in terms of welfare to a time of thinking of the developmental stage of "youth" as perhaps as no longer necessary and the object of punishment:

> The era of "child-saving" in the United States ended with welfare "reform" in 1996. The resources once committed to education, health, and social welfare programs of panoptically viewed youth and children are now utilized to build prisons, install metal detectors in schools, and criminalize younger children as adults. (2001, p. 198)

From this point of view, salvation in NCLB can be interpreted as a technology for fast-tracking children to the final state of adulthood. Reading and math skills will more quickly prepare children to take their place as independent, working citizens in the marketplace.

All of these constructions of the individual are examples of current discourses that examine the relations of power/knowledge that construct the child differently, exposing the dangers that accompany any reform. They all analyze the inscription of power/knowledge on the body, structuring the possibilities and impossibilities of what it means to be a child, and what it means to govern oneself and be governed. The shifting meanings of the child open the possibilities of thinking salvation, reform, and the child differently.

New Patterns of Salvation

The new patterns of governance in the reform of NCLB join individual competence, accountability, and responsibility with laws and norms in new networks of power and knowledge in the salvation of the child. One can read these transformations as a restructuring of biopower, or the administration of life, both at the level of individual bodies and at the level of populations. The health and growth of the social body requires the reinforcement of both of these factors (Foucault, 1976/1990, pp. 141–145). NCLB, in the ongoing pattern of redirecting forces from welfare to work, utilizes technologies of sanctions and targets of power to shape a docile but productive population. The power of law is embedded in the new norms of proficiency, accountability, choice, and flexibility. Just as charity as a cure for poverty was found wanting in the seventeenth century with the giving of alms in exchange for salvation, and in the nineteenth century with philanthropy as a secular salvation, welfare was found wanting in the late twentieth century. It was a burden on the health of society. The strength of society depends on the productiveness of the evolving child, and this capability is defined and measured by NCLB as instilling reading and math skills into the body of the child.

When one examines the conceptual relations of power and knowledge in salvation, one finds many different constructions. The simple purchase of alms or philanthropy is one; there is in the language of the legislation a sense of "belief" and a measure of "giving" to the needy. President George W. Bush was quoted saying "These reforms express my deep belief in our public schools and their mission to build the mind and character of every child, from every background, in every part of America," and further "President Bush emphasized his deep belief in our public schools, but an even greater concern that "too many of our neediest children are being left behind" (U.S. Department of Education, 2002, ¶1).

Often, the allusions to salvation[4] in educational literature are in the language of Protestant Calvinism. There are, however, many transformations of Calvinism, especially during the eighteenth century. For the purposes here, I will briefly utilize three patterns, which span a trajectory through the American Enlightenment (Haroutunian, 1964). The first pattern, represented by Jonathan Edwards, was the traditional construction which postulated intense original sin, the need to achieve a "regenerate state" through work on the self as the passage to "everlasting life" and the damnation to the miseries of everlasting hell for all the unregenerate souls. Salvation was revealed. The second pattern, represented by Charles Chauncy, downplayed the intensity of original sin and postulated the need to achieve a regenerate state for everlasting life only in relation to the degree of one's sin. He recognized multiple causes and effects of degeneracy that evolved into delinquency and the diversity of obligations and responsibilities. Damnation to hell was only a temporary state according to one's degree of sin. Salvation was "rescue." Eventually all souls would find their way to heaven. The third pattern, considered the idea of original sin as "barbarism." Sin was of this world, God willed the happiness of all, repentance was retranslated as reform, right conduct became a matter of ethics and law, and, finally, there was no need for hell because all souls were already saved by love and grace (Haroutunian, 1964).

The various ideas of salvation have been transformed and secularized, but strands of these ideas can be found recirculated and incorporated into the heterogeneous strategies of NCLB. For example, the idea of salvation for all in its title, *No Child Left Behind*, resonates with the third form which implies on the surface, at least, that life is already saved. On closer analysis, however, this is not so certain. Children must perform to receive salvation, and further, they must perform in the same way. NCLB assumes a unitary and universalized standard, which is more homologous with the first form of salvation than the third. There is no choice, or flexibility, in the manner suggested in patterns two and three; there is only accountability.

The sanctions in NCLB suggest at times outright damnation, as "persistently failing schools" and "trapped in a failing school," and sometimes to simply disappear in the "risk of reconstitution." Sometimes there is the opportunity to earn one's way out of "hell" through corrective action like supplemental educational services to eligible students (i.e., those destined for or condemned to hell). In its language, NCLB seems measured in its punishments. In either case, there is clearly the idea of the need for . redemption. In the following example, we have the evidence of total condemnation. As of December 2002, Deborah Lynch, president of the Chicago Teachers Union, said that three schools in Chicago were scheduled for closure for chronic academic failure; "fourteen hundred poor minority children were displaced from their neighborhood schools" (Speingen, 2002, p. 74). In its effects, NCLB may appear more as a permanent, everlasting hell.

There are other effects of power and knowledge in the construction of the child. As another effect of NCLB, Wisconsin's Department of Public Instruction (WDPI) held a series of public hearings on its proposed definition of a "persistently dangerous" school. Expulsion data is the focus of WDPI, but there are unintended consequences of NCLB because schools can control the number by deferments and deals by manipulating meaning. David Schmidt, Waukesha School District Superintendent, said, "They defer, meaning they say to people 'You'll do this, this, this, and then you can continue to stay in school.' Or they cut a deal and do voluntary withdrawals. A lot of schools do that to avoid expulsions" (*Oshkosh Northwestern*, 2003). NCLB makes visible Foucault's (1984) warning of danger in the choices of the constructions of the child, "My point is not that everything is bad, but that everything is dangerous, which is not exactly the same as bad. If everything is dangerous, then we always have something to do. So my position leads not to apathy but to a hyper- and pessimistic activism" (p. 343).

Another effect of power in NCLB is the reiteration of "the need to learn to learn again." There is a continuous pattern of testing and retesting in a way that is reminiscent of the interrogation of the soul. There is no assurance of being saved. It has often been observed that people need to be enterprising; they must learn to be flexible, classifiable, mobile, declassifiable, that is "to learn and to learn to learn again" (Granel, 1991, p. 154). Power requires this constant vigilance to stay in a state of grace. This injunction resonates with the traditional view of Calvinism which requires individuals to constantly revisit their sins and to confess them and confess again and again: "Be often confessing your old sins to God, and let that text be often in your mind" (Edwards, 1741/1830, p. 151). One never escapes the effects of sin and there is no assurance of receiving redemption. This resonates with the first pattern of salvation that I discussed earlier. There is no such resonance with the other two patterns of salvation.

In NCLB, there in an increase of dependence. The child must stay close to the law and administration; they must submit to the Secretary of Education in the performance agreements. Salvation for the child is a continuous walk through the perils of educational hoops of measurement and division. Jonathan Edwards (1741/1830) gave parallel advice of dependence to a young parishioner when she asked him how to achieve salvation. Point 16 of his response reads, "In all your course, walk with God, and follow Christ, as a little, poor, helpless child, taking hold of Christ's hand, keeping your eye on the marks of the wounds in his hands and side, whence came the blood that cleanses you from sin, and hiding your nakedness under the skirt of the white shining robes of his righteousness" (p. 152). The NCLB child must always stay close to the mark.

These examples of relations of power and knowledge and the effects of power in the construction of the child point to a heterogeneous restructuring of strategies and practices that make present constructions of the child possible. They also point to the dangers not only identified within the discourses of reform as "dangerous schools," "dangerous populations," or "at risk" but also the dangers of the effects of power. These dangers, often veiled and ignored in the pursuit of particular stated outcomes, are consequences of blind, heterogeneous strategies of reform.

Summary

Understanding salvation as analogous to the construction and education of the child is a heterogeneous and dangerous undertaking. There are many forms of salvation and we need to be cautious in the application of the term at the intersection of the construction of the child and education. The invocation of the term "salvation" suspends a blanket of faith over the strategies and technologies that are the particular practices of education. It veils and obfuscates meanings. It is important to interrogate the effects of the "feeling" with which religious discourses blanket meanings. The effects of the discourses of salvation disguise and push to the shadows how subjectivity is produced and how the relations of power and knowledge circulate in the production of the lives of children and their families.

Reforms are never entirely new. I have pointed to some of the dangers that travel the lines that intersect the constructions of the child, salvation, and education in NCLB. The technologies and practices are often reworked and restructured, practices that historically have circulated again and again. A prominent example is the recurrence of the effort to construct a universalized concept of the child where all children are everywhere the same. NCLB is no exception. The strategies and technologies which structure its heterogeneous, conflicting patterns, are visible in other historical moments. By bringing history to the fore, some of the effects and harm of unintended consequences of reform can be anticipated and give direction to further questions.

Notes

1. An earlier version of this chapter was presented at the American Educational Research Association in Chicago, Illinois on April 22, 2003.
2. It is not unusual to construct or reshape a concept by drawing on meanings that flow from dissimilar and mixed discourses. This is one meaning of heterogeneity. The mixture of meanings can construct concepts that upon analysis are inherently contradictory, with tensions that make unified meanings of concepts, such as "the child," impossible. For example, in this chapter the heterogeneity of the concept "salvation" renders a proliferation of the meanings of NCLB. Homogeneity is likewise troublesome. Trying to hammer all possible meanings of the child into one mold to "produce" a unified subject that moves in lock-step through a graded school curriculum is equally impossible. The mechanical assembly line production of the child is a fantasy that has circulated as a model for education since the creation of the assembly line. The periodic testing of the child in school is a direct transformation of the idea of quality control in the factory. A common assumption is that if quality control produces a uniform, homologous product coming off the assembly line in the factory, the same practices should surely produce the same sameness in children, where all are of equal value, structure, and function.
3. Meaning refers to the construction of knowledge, not as universalized ideas, but rather as a gap where understandings are in perpetual oscillation (see Foucault. 1994, chapter 10).
4. Historically, there is a transformation from thinking three realms of existence (mind, body, and soul) to thinking two realms of existence (mind and body). The soul, which received so much attention throughout medieval times, fades and the mind emerges as the site of attention in the Enlightenment. During the twentieth century, the mind faded and the body emerged as the site of attention. Some writers today prefer to think of one realm of mind–body. Also, today there seems to be a reemergence of a third realm, no longer as soul, but a "spiritual life."

References

Aries, P. (1964). *Centuries of childhood: A social history of family life*. New York: Alfred A.Knopf. (Original work published 1962)

Baker, B. (1998). "Childhood" in the emergence and spread of the U.S. public school. In T.S. Popkewitz & M. Brennan, (Eds.), *Foucault's challenge: Discourse, knowledge, and power in education* (pp. 117–143). New York: Teachers College Press.

Bourdieu, P. (2000). *Pascalian meditation*. Stanford, CA: Stanford University Press. (Original work published in 1997)

Cremin, L. (1961). *The transformation of the school: Progressivism in American education,* New York: Vintage Books.

Derrida, J. (1967/1976). *Of grammatology*. Baltimore, MD: The The Johns Hopkins University Press.

Donzelot, J. (1997). *The policing of families*. Baltimore, MD: The The Johns Hopkins University Press. (Original work published 1977)

Dussel, I. (2002). Educational restructuring in Argentina: Hybridity, diversity, and governance after welfare. Paper presented at the Symposium: Controversies in education restructuring in Lisbon, September 11–14.

Edwards, J. (1830). *Works of President Edwards*. (New York: G&C&H Carvill). (Original work published 1741)

Foucault, M. (1972). *The archeology of knowledge and the discourse on language*. New York: Pantheon Books.

Foucault, M. (1977). *Language, counter-memory, practice: Selected essays and interviews*. Ithaca, NY: Cornell University Press.

Foucault, M. (1980). *Power/knowledge: Selected interviews and other writings 1972 – 1977*. New York: Pantheon Books

Foucault, M. (1984). *The Foucault reader*. New York: Pantheon Books.

Foucault, M. (1988). *Politics, philosophy culture: Interviews and other writings 1977–1984*. New York: Routledge.

Foucault, M. (1990). *The history of sexuality: An introduction* (Vol.1). New York: Vintage Books. (Original work published 1976)

Foucault, M. (1994). *The order of things: An archaeology of the Human Sciences*. New York: Vintage Books. (Original work published 1966)

Franklin, B. (1961). *The autobiography and other writings*. Berkeley, CA: Signet Classic.

Granel, G. (1991). Who comes after the subject? In E. Cadava, P. Connor, & J. Nancy (Eds.), *Who comes after the subject?* (pp. 148–156). New York: Routledge.

Hacking, I. (1995). *Rewriting the soul: Multiple personality and the sciences of memory*. Princeton: Princeton University Press.

Hall, G. S. (1908). *Adolescence*. New York: Arno Press and The New York Times. (Original work published 1904)

Haroutunian, J. (1964). *Piety versus moralism: The passing of the New England theology*. Hamden: Archon Books.

Hunter, I. (1992). Aesthetics and cultural studies. In L. Grossberg, C. Nelson, & P. Treichler (Eds.), *Cultural studies* (pp. 347–372). New York: Routledge.

Lesko, N. (2001). *Act your age! A cultural construction of adolescence*. New York: Routledge Falmer.

Locke, J. (1961). *An essay concerning human understanding*. London: Everyman's Library. (Original work published 1690)

Locke, J. (1996). *Some thoughts concerning education*. Indianapolis, IN: Hackett. (Original work published 1693)

O'Donnell, J. (1985). *The origins of behaviorism: American psychology, 1870–1920*. New York: New York University Press.

Popkewitz, T. S. (1998). *Struggling for the soul: The politics of schooling and the construction of the teacher*. New York: Teachers College Press.

Rose, N. (1990). *Governing the soul: The shaping of the private self*. London: Routledge.

194 *Pena*

Schmidt, D. (2003, February 31). DPI to hold public hearings on dangerous schools definition: State is required to come up with a list of dangerous schools by next summer. *Oshkosh Northwestern*, p. C4.

Speingen, K. (2002, December 30). Reformer in the ranks. *Newsweek*, Vol. 141(1) p. 74.

U.S. Department of Education (2002). *The No Child Left Behind Act of 2001: Executive Summary*, Washington, D.C. Retrieved November 7, 2002, from http://www.ed.gov/nclb

U.S. Department of Labor (June 1991). *What Work Requires of Schools: A SCANS Report for America 2000*, Washington, D.C, June.

CHAPTER 10

Illusions of Social Democracy: Early Childhood Educational Voucher Policies in Taiwan

I-FANG LEE

Introduction

Circulating and traveling around the globe, vouchers and school choice discourses currently function as the new "truth" for educators and parents, mobilizing them to imagine different ways of changing the field of education for the better. Without thorough critical investigations/inquiries, educational voucher policies are often thought of as examples of democratic educational reforms in multiple continents, including South America, North America, Australia, Asia, and Europe.[1] Supported by Milton Friedman's discussions on educational vouchers as tactics to dismantle government's monopoly over modern public schooling systems and mobilized by neoliberal discourses of socioeconomic and cultural reform discourses, notions of educational vouchers and choice discourses are packed with ideas of choice and promises of social democracy for the field of education.

My efforts to deconstruct the mentalities which sustain vouchers and school choice discourses, and the "regime of truth" that is produced through voucher policies are motivated by a Derridian notion of ethical attitude (Derrida, 1988). By invoking Derrida's notion of deconstruction, I intend to unpack the educational idioms and discourses surrounding contemporary educational voucher policies to open up new spaces for a rethinking and reenvisioning of how new ways and forms of sociocultural governances are produced through the languages of educational reform policies. Therefore, with ethical concerns about issues of social justice, I argue that voucher policies are not "magical" solutions to "fix" educational problems but instead produce illusions of social democracy by upholding concepts of "freedom to choose" and "autonomous entrepreneurial selves" (Rose, 1998) as dangerous new "truths." To understand the effects of educational vouchers better, this chapter will focus on the following questions: (1) What are the conditions and mentalities that allow the deployment of educational vouchers in Taiwan? (2) How do educational voucher policies discipline the field of early childhood education and parents/families simultaneously? And (3) what kinds of new "norms" or "truths" are produced through the deployment of early childhood educational voucher policies in Taiwan?

My interest in reconceptualizing the effects of educational voucher policies is not to discuss whether vouchers are good or bad for dismantling government's monopoly over public school systems. Instead, I intend to examine how early childhood educational voucher policies in Taiwan produce new ways of thinking about what "modern and democratic" educational practices are and how "modern and well-educated" autonomous individuals should act and think.

In this chapter, voucher policies will be conceptualized as sociocultural governing technologies that work to construct, shape, and "normalize" new desirable "norms" of being, acting, and thinking within the field of education. Understanding educational reform policies (such as vouchers) as sociocultural governing technologies requires an analytical shift, which embodies a linguistic turn in order to understand notions of power and investigate how particular new "norms" are produced through the languages of educational reforms and discourses. In addition, rather than a repressive notion of power, which views government/the state as the sovereign entity with power to direct changes in the field of education, a productive notion of power will be employed to open up a new discursive space for us to interrogate how new systems of reasoning are created through the circulation of vouchers and through school choice discourses. Such a theoretical change in analyzing notions of power in order to understand the effects of voucher policies destabilizes popular political rhetoric and critical discussions on educational vouchers as "magical" reform policies toward a modernized field of education with equality and social democracy.

In the following sections of this chapter, I will start by historicizing the field of early childhood education and care in Taiwan. By historicizing the present, I aim to trace the shifting and changing cultural definitions of "appropriate" early educational and care programs for young children across different historical moments. This method underscores how new reasoning/knowledge systems are crafted to scaffold particular mentalities and conditions for the deployment of educational voucher policies. Historicizing the present treats the field of early childhood education and care as a discursive space to highlight how multiple tensions, forces and competing notions of "appropriate" early educational and care programs for young children are circulating against each other and/or work hand-in-hand to fabricate particular conditions/mentalities to allow the deployment of voucher policies.

In the second section of this chapter, the deployments of voucher policies in Taiwan are conceptualized as amalgamations of global circulations of vouchers and school choice discourses. These discourses incorporate and produce a particular national imaginary of "modernity" and "democracy" specific to this historical time and space. After historicizing the specific historical, political, and sociocultural conditions for the deployments of early childhood educational voucher policies in Taiwan, I interrogate how vouchers produce new ways of thinking/reasoning about what are modern and democratic early educational and care programs and how parents/families should act/respond/choose.

In the third section of this chapter, through poststructural theoretical analytic concepts, such as Foucault's (1991) notions of power/knowledge and governmentality, and Popkewitz's (2000a) idea of indigenous foreigner and hybridization, I problematize the new "truths" and "norms" that are produced by voucher policies/educational reform policies in Taiwan. Recognizing the critiques of critical structural analyses concerning class struggles and social inequalities through the deployments of voucher policies (Apple, 2000; Carnoy and McEwan, 2003), the third section of this chapter illustrates how vouchers and school choice discourses create new and dangerous "universal" narratives for ways of being and acting. I argue that as the new image of modernized "autonomous entrepreneurial selves" (Rose, 1998) becomes the desirable model/norm for all parents and families, critical structural binary/category discussions of social inclusion/ exclusion will be limited/restrictive. Instead, the issue of social inclusion/ exclusion associated with voucher policies needs to be thought of as a notion of relations and/or a map-making concept that simultaneously includes particular groups of parents/families with certain sociocultural dispositions as the "norm" while excluding others as needing to be "reformed" or "modernized." The poststructural theoretical shifts and organization of the three sections within this chapter are not undertaken with the intention of denying the importance of critical theoretical analyses of voucher policies. Rather, through historicizing the changing meanings of "appropriate" and "modernized" early educational and care programs, deconstructing the deployment of voucher policies, and re-conceptualizing the new "norms/truths" that are produced through the deployment of voucher policies, I argue that educational reform policies such as vouchers are illusions of social democracy.

A History of the Present: Changing Meanings of
Early Childhood Education and Care

A history of the present destabilizes our "natural" ways of reasoning and ways of being. As Popkewitz *et al.* (2001) assert in their work:

> The history of the present aims to grasp the conditions concerning what is possible to say as "true," and to consider the present configuration and organization of knowledge through excavating the shifting formations of knowledge over time (p. 32).

As I look at the changing meanings of early childhood education across different historical times, the notion of history is not treated as a linear timeline but rather as a reconceptualization of the present. In the following section I illustrate how early childhood education and care has come to be defined and thought of as it is currently and how "modern" and contemporary constructions of "appropriate" early childhood education and care programs produce new conditions/mentalities for the deployment of voucher policies in Taiwan.

Early Childhood Education and Care in Taiwan

(1940s to the Present)

Since the 1940s,[2] an array of preschool programs such as community/village childcare facilities during agriculture harvest seasons, church- or temple-based early childhood programs, and various childcare provisions for parents who are in the military have dominated the field of early childhood education. Among these different types of programs for young children, distinctions (such as private versus public) and organizations (such as kindergarten programs or child care institutions) of these institutionalized early educational programs were left unregulated until the first Preschool Education Act was put into place 1981.[3] The term "kindergarten" was used loosely as an overarching label to describe all kinds of institutionalized programs for young children in Taiwan prior to 1981.[4] In addition, distinctions between public and private early educational and care programs were also blurry because public funding for the field of early childhood education and care was neither budgeted nor systematically allocated and distributed. Before the Preschool Education Act in 1981, programs for young children functioned as transitional child care programs for parents prior to the child's entry into compulsory education at age seven.

The appearance of the Preschool Education Act in 1981 was a response to prevailing critiques by multiple early childhood educational researchers and development discourses concerning how "un-scientific" and "un-systematic" the field was when compared with early childhood education and care systems of multiple "well-developed" nations from the West. To emulate what reformers considered "advanced" and "modernized" early childhood education and care systems from the West (mostly European countries and the United States, as well as Japan). These were considered necessary to reform and restructure the field of early childhood education in Taiwan. Thus, official distinctions of education/care and private/public for the multiple forms of early childhood programs were constructed through the Preschool Education Act in 1981. These official distinctions were deemed necessary and desirable for the field of early childhood education and care to become modernized and well organized in Taiwan.

As desired and intended, "modern" distinctions of public versus private and "scientific" organizations of the kindergarten versus childcare programs within the field of early childhood education and care are produced through the deployment of this early childhood educational reform policy in 1981. The texts and languages of this educational reform policy not only functioned as modernizing reform policies but also produced new mentalities for thinking about what "modernization" means or entails within the field of early childhood education and care in Taiwan. Along with the deployment of this early childhood educational reform policy from 1981 were new discursive spaces for imagining what modernization means within the field of early childhood education and care. The enactment of this educational reform policy was recognized as a "milestone" (Lu, 1984). This legitimized early childhood education as a unique field for young children's early years of education. Among the original 25 points of the Preschool Education Act in 1981[5] are several noteworthy points which illustrate the shifting meanings of "appropriate" and desirable early educational and care programs for young children. For example:

Point 1: The main purpose of early education is to facilitate a child's physical and psychological well-being and development . . .
    Point 4: Kindergarten programs that are operated by municipalities or counties and/or are affiliated with teacher education programs of public universities and public elementary schools will be classified as public institutions. Other kindergarten programs that are not affiliated with any public/government budget will be classified as private institutions.[6]

Through this educational policy, programs that focused on aspects of "education" were called kindergarten programs to be regulated by the Ministry of Education. Programs that focused on notions of "care" were classified as childcare institutions to be governed by the Children's Bureau in the Ministry of the Interior. In addition, programs with public/government funds became identified as public institutions, while programs with private money/budgets were categorized as private institutions.

These official differentiations of public/private and kindergarten/child care changed the meanings of early childhood education and care and created new discursive spaces and new sociocultural conditions for multiple actors within the field of early childhood education and care. This reformulated space made it possible to imagine different ways of constructing new connotations of "ideal" and "good" early childhood education and care programs. Since 1981, contemporary constructions of "ideal" and "good" early childhood education and care programs in Taiwan have shifted and changed to define early childhood as a critical period of early learning, development, and growth. Thus, contemporary kindergarten programs or child care institutions in Taiwan have started to emphasize how to better facilitate young children's cognitive growth and sociocultural development before their entrance into compulsory education. Various international early childhood development theories and pedagogies translated from the writings of scholars such as Piaget, Vygotsky, Froebel, and Montessori are popular contemporary methodologies that have recently traveled into the field of early childhood education and care in Taiwan. The new definitions of "modernized" and/or "ideal/good" early childhood education and care programs also incorporated Chinese/Taiwanese cultural understandings of the "educated" child. They reinforced "Western" concepts of the child but also became a new "regime of truth" that shapes contemporary construction of the "educated" and "modern" child in Taiwan.

Evidently, a particular notion of what are considered "modern" and desirable directions of development in the field of early childhood education and care in Taiwan reflects a history of power/knowledge relations that are associated with the production of a geopolitical space (Escobar, 1995) of power and the fabrication of a particular contemporary national imaginary. When reading the Preschool Education Act of 1981, it becomes evident that contemporary constructions of modernization in Taiwan incorporate ideas of Western standards of progress, development, and modernity as the "truth" or model. Such imaginaries and translations of modernization indicate a linear notion of what development is, as Western experiences of civilizations are upheld as desirable "models" to be followed in Taiwan. Using the "West" as the standard for what is considered to be "modern" and "well-developed" is dangerous, as critiqued in Escobar's work on the singularity of the "Western" notion of development.

> I propose to speak of development as a historically singular experience, the creation of a domain of thought and action, by analyzing the characteristics and interrelations of the three axes that define it: the forms of knowledge that refer to it and through which it comes into being and is elaborated into objects, concepts, theories, and the like; the system of power that regulates its practice; and the forms of subjectivity fostered by this discourse, those through which people come to recognize themselves as developed or underdeveloped. (1995, p.10)

Escobar's analysis of the forms of subjectivity speaks to contemporary constructions of the national imaginary in Taiwan. Present constructions of "modern" early childhood educational programs in Taiwan are supported by development discourses and cultivated by a national imaginary of what it understood to be "modern" that reflect the "systematic," organized early educational and care programs in the "modernized" and "well-developed" Western nations.[7]

Within the field of education, the supposed "modern" and/or progressive educational reform practices from the "West" are commonly identified as the discourses of "parental choice," "high quality programs," "scientific curriculum," "equality for all" (Bloch, 2003), and so on. While each discursive notion embodies different political ideologies, pedagogical practice, and theoretical ideas, these ideas function as educational reform slogans and travel together into the field of early childhood education in Taiwan as desirable "modern" and/or progressive practices. The amalgamation of a national imaginary of "modernization" in Taiwan, global circulations of "Western modernity" as the "norm" and Western educational reform experiences/discourses as the new "truth" work together to foster particular conditions for the formation and deployment of early childhood educational vouchers in Taiwan.

Contemporary Formation and Deployment of

Early Childhood Educational Voucher

Policies in Taiwan

Early childhood educational voucher discourses in Taiwan are strongly connected with new cultural constructions of the "modern well-educated" child. This "modern" child is envisioned through a national imaginary of modernization. The deployment of the first early childhood educational reform policy in 1981 in Taipei ruptured the field of early childhood education and care and created a new sociocultural condition within the field, as the distinctions between education/care and public/private were legitimatized. Associated with these categories of public/private, education/care are new educational and social issues concerning fairness, equality, and quality within the field of early childhood education and care.

Recognizing early childhood as a critical period for growth and development, early childhood educational and care programs are becoming important characteristics of an early head start for young children in Taiwan. According to a parental survey in Taipei (Association of Early Childhood Education in Taiwan, 1998), nearly 96 percent of five-year-old children, 91 percent of four-year-old children, and about 60 percent of three-year-old children are reported to be enrolled in either public or private kindergarten programs and childcare institutions.[8] As the numbers of public and private kindergartens and child care programs have increased to echo the importance of early childhood education and care, it is important to note that private institutions dramatically outnumber public programs. According to the statistical records from the Ministry of Education and the Children's Bureau during the 2000–2001 school year, there were 7.2 times more private kindergartens than public ones and 6.9 times more private child care programs than public ones. By taking such dramatic differences between the numbers of public and private programs in the 2000–2001 school year as an example, and connecting this example with the seemingly high percentage of attendance/enrollment in kindergarten and child care institutions, it becomes possible to say that the accessibility of public programs for young children is limited in Taiwan. This problem of inadequate numbers of public programs for all young children is further perpetuated through a lottery enrollment method.

Currently, enrollment for young children in public early childhood education and care programs is organized through a lottery system which operates on a district-by-district basis. Such an enrollment method requires parents who desire to enroll their children in public programs (both kindergartens and child care institutions) to either know or learn how to maneuver the rules of a lottery game.[9] If parents cannot win spots in public programs for their children through the lottery system they have no choice but to look into private programs since the period of early childhood has been constructed and deemed a critical period for learning and growth. At this point in the discussion, it is also important to draw attention to tuition differences between public and private programs in Taiwan. Taking Taipei (one of the most expensive cities in which to reside in Taiwan) as an example, the average tuition for private kindergartens is 2.8 times greater than that of public ones (Taipei City Government, 1997).[10]

The lack of public programs and the high tuition cost of private institutions for young children have increased the desire for further reform of the field of early childhood education and care in Taiwan. One of the prevalent arguments highlighting the lack of equality (fair public program enrollment) within the field of early childhood education and care is a combination of critiques of the current lottery enrollment method and dissatisfaction with how the central government/state is investing more money in children attending public programs.

Accompanied by global circulation of voucher and school choice discourses, parents with young children in private programs and educators from multiple backgrounds have formed coalitions to further the idea of educational vouchers as a progressive, democratic, and "modernized" educational reform practice (Lu, 2001; Lu and Hsieh, 2001; Lu and Shih, 2000). In addition, parents with children in private programs have argued that since all parents are taxpayers, they are being double-dipped by the government/state because the number of public programs is not adequate, nor are they accessible to all children.

Attempting to address such dissatisfactions within the field of early childhood education and care, in 1994, Mr. Shui-bian Chen,[11] the Democratic Progressive Party candidate for mayor in Taipei, promoted early childhood educational vouchers as a way to increase young children's access to early educational programs, to support parental rights to choose their children's educational programs, to facilitate positive competition in the field of early childhood education for higher quality, and to encourage nonlicensed programs to become licensed. In addition to Mr. Chen's political campaign promise, in 1998, multiple groups of parents and educational researchers formed an alliance with the Early Childhood Education Association of the Republic of China (ECEAC) to organize a social demonstration concerning issues of fair (re)distributions of public funding/resources in the field of early childhood education and care. This social demonstration, known as A Walk for Early Childhood Education, brought 1,018 parents and educators onto the streets of Taipei to demand more government funding/budget for early childhood education. Among the multiple requests from this group of parents and educators, the idea of the early childhood educational voucher has been mobilized and publicized as a form of "social justice" (Pan, 2000).

Hence, toward the end of Mr. Shui-bian Chen's term (1994–1998) as mayor of Taipei City, in August of 1998, the Education Bureau of Taipei City allocated funds for vouchers in its annual budget. This first budget move legitimized the first early childhood educational voucher policy for families/children who are eligible and qualified in the 1998–1999 school year and to foster the island wide deployment of voucher policies in Taiwan.[12] Since August 1998, many cities and counties in Taiwan have gradually followed such a reform policy, and in the 2000 school year,[13] the voucher reforms/policies were institutionalized islandwide with equal amounts of money.[14] While early childhood educational voucher policies were enacted in different years, the rules of such policies have been identical. Currently, throughout Taiwan, early childhood educational vouchers are for five-year-olds attending licensed private kindergarten programs or child care institutions.

The formation and deployment of early childhood educational voucher policies in Taiwan is not accidental but is scaffolded by multiple discourses concerning how to better change the field of early childhood education and care toward modernization. Early childhood educational vouchers in Taiwan are recognized and constructed as progressive reform policies. This change is fostered by the globalization of voucher and choice discourses and fueled by dissatisfaction with current unequal government investment between children in public and private programs, and accompanied by the contemporary construction of a national imaginary of what it means to be "modern" and "educated" child. In addition, early childhood educational vouchers are deemed to be a "social justice that comes late" (Pan, 2000) for parents and children in Taiwan. Therefore, they are embedded within educational voucher policies as mythologies of "freedom to choose" and illusions of social democracy.

Deconstructing and Rethinking Early Childhood Educational Voucher Policies in Taiwan

While educational vouchers have become a globalized discourse instilling notions of freedom, choice, and a promise of democracy to come, opponents of vouchers have categorized such reform policies as advocating neoliberal ideologies (Apple, 2000; Cookson, 1994). Discussions on vouchers as neoliberal reforms that further perpetuate social inequalities stem from critical traditions and do provide thoughtful analyses on issues concerning class struggles and social exclusions. While critical analyses of vouchers are vital to help us understand vouchers and choice discourses as dangerous "othering practices" that exclude children and families, further deconstruction of how voucher discourses produce new "norms" and/or new "truth" heighten our awareness of how reform policies function as sociocultural governing technologies to shape and construct new "modern norms" of reasoning, acting, and being.

To debunk the myth of voucher policies and choice discourse as new democratic educational reform policies, vouchers can be thought of as a form of "planetspeak" (Nóvoa, 2002; Popkewitz, 2003) that circulates globally forming "universal truths" or a "worldwide bible" advancing solutions to all problems within the field of education. In other words, through vouchers, ideas of freedom to choose and notions of fairness/social justice within the field of early childhood education and care are all mobilized as desirable "democratic norms."

At this point, I turn to a discussion in which contemporary educational reform discourses, such as voucher policies in Taiwan, can and should be thought of as hybrids of globalization/localization discourses that weave both global and local systems of knowledge and mentalities through the deployment of the notions of hybridization and indigenous foreigner explained by Popkewitz (2000a).

The concept of *hybridization* makes it possible to think of educational practices as having plural assumptions, orientations, and procedures.

> National reform practices, as well, can be understood as practices that have multiple assumptions and divisions from which the political imaginary is being revisioned. The globalization of educational reforms embodies a complex scaffolding of techniques and knowledge that are not imposed through fixed strategies and hierarchical applications of power that move uncontested from the center nations of the world system to the peripheral and "less powerful" countries." (2000a, p. 272)

Educational voucher policies in Taiwan embody combinations of decentralization and Chinese/Taiwanese cultural reasoning of "modernized" early educational and care programs to form a distinctive set of sociocultural governing practices. By remixing global notions of choice and decentralization with local constructions of "modern" educational practices, contemporary early childhood educational vouchers policies in Taiwan have amalgamated global educational reform discourses with local national imaginary of modernization to shape and fashion desirable educational practices in Taiwan.

Being global and local simultaneously, contemporary educational reforms such as educational voucher policies in Taiwan produce sociocultural governing practices that can be understood through Popkewitz's (2000a) concept of the indigenous foreigner. As it is explained:

> It is common in national policy and research for the heroes of progress to be foreigners who are immortalized in the reform efforts. The names of the foreign authors, for example, appear as signs of social, political and educational progress in the national debates . . . While the heroes and heroines circulate as part of global discourses of reform, such heroes and heroines are promoted in national debates as indigenous in what appears to be a seamless movement between the global and the local. The foreign names or concepts no longer exist as outsiders but as indigenous without alien qualities. The invocation of the

indigenous foreigner functions to bless the social reform with the harbinger of progress. (2000a, pp. 277–278)

Examples of the indigenous foreigner through contemporary educational reform discourses, to name a few, have been about decentralization, marketization, freedom to choose, quality and standard movements, and deregulation/reregualtion in the field of early childhood education in Taiwan. Debates on deploying early childhood educational vouchers indicate the hybridization of local conceptualization or (re)appropriation of global and foreign concepts in Taiwan. Intertwined with the notion of indigenous foreigner is a complex web of power/knowledge relations in which certain actors (such as educational researchers) are perceived to have the cultural capital (Bourdieu, 1984) or authority to mobilize and indigenize certain global discourses for the benefit of all.

Thus, the concepts of indigenous foreigner and hybridity not only recognize how global and local reform discourses overlap but also refuse to accept educational voucher policies as universal sets of ideas/rhetoric. In addition, the importance of the concept of the indigenous foreigner is not the foreign concepts or names that are immortalized but rather the hybridity of the discourses that legitimate the forms of knowledge and experiences of modernity and social progress as "regimes of truth." These concepts travel across space and time but in historically contingent ways (Popkewitz, 2000a).

Recognizing the formation of voucher policies in Taiwan as mixtures of global circulations of educational reform discourses and local desires to modernize educational practices, a Foucauldian reading of educational voucher policies will interject new understanding of how languages and texts of reform policies work to construct and legitimatize new "norms."

A Foucauldian Reading of Early Childhood
Educational Voucher Policies in Taiwan

To destabilize and interrogate common understandings, mentalities, and prevalent discourses that have worked together to scaffold the myth of early childhood educational voucher policies as progressive changes for greater social inclusion in Taiwan, I use Foucault's (1991) notion of governmentality to move away from a sovereign repressive notion of power. Instead of positing power as a thing to be redistributed by the power holder (such as the sovereign state), the notion of governmentality shifts to focus on the productive power of voucher policies to understand how vouchers and choice discourses discipline and construct new mentalities within the field of education.

In addition to discussions concerning questions of how to govern oneself, how to govern others, and/or how to be governed, contemporary understandings of the art of government center on how to practice "democratic" modes of governing for a future imagined to be better and more prosperous. Through the deployment of voucher policies and choice discourse not only are the early childhood educational and childcare programs being governed but parents or families are also being administered. To better understand how vouchers and school choice discourses function as modern democratic technologies of dual-governance for both parents and the field of early childhood education and care in Taiwan without the need to deploy brutal force, I use Foucault's interpretations of power/knowledge to illustrate how the language of policies fabricates and/or inscribes new ways of thinking about who is and who is not the new "modern educated subject." Interlaced with a national imagination of modernity and development, voucher and choice discourses function as modern democratic governing technologies to produce a new "regime of truth." As Foucault (1980) notes:

> "Truth" is to be understood as a system of ordered procedures for the pro-
> duction, regulation, distribution, circulation, and operation of statements.
> "Truth" is linked in a circular relation with systems of power which pro-
> duce and sustain it, and to effects of power which it induces and which
> extend it (p. 133).

Within this poststructural theoretical framework of understanding, educational voucher policies can be thought of as sociocultural governance practices that aim to (re)shape, regulate, and construct what are thought to be reasonable and desirable modes of thinking and acting (Bloch & Blessing, 2000; Dahlberg, 2000; Lather, 2004; Popkewitz, 2000b; Rose, 1998). These new norms are productive in the sense that parents and children are transformed to become responsible, autonomous, and enterprising individuals (Dahlberg, 2000; Rose, 1999).

Shore and Wright's (1997) work envisioning how policies function as instruments of governance also enables us to understand how parents become so-called "entrepreneurial individuals" (also see Dahlberg, 2000; Rose, 1998) when exercising their "parental choice." Viewing policies as "anthropological phenomena" through a Foucauldian lens creates a space of possibilities to see how vouchers work to discipline families through normalizing practices that function to include families from certain social spaces, while at the same time excluding other families. Thus, early childhood educational voucher policies in Taiwan function as a sociocultural governing practice. This practice regulates the field of early childhood education and (re)envisions the field as a "quasi-market" with notions of "free choice" as parents are disciplined and reconstructed to become entrepreneurs and consumers (Apple, 2001; Gewirtz *et al.*, 1995; Rose, 1998).

Early Childhood Educational Voucher Policy
as a Socio-Cultural Governing Practice

The guidelines and rules that underpin early childhood educational voucher policy function as a mode of sociocultural discipline to shape a new sociocultural understanding of what early childhood education means for children, parents, and teachers. The notion of parents as entrepreneurial selves or consumers within early childhood educational "markets" also embodies notions of the "appropriate" types of early educational and childcare programs and how to choose "appropriate" early educational and care programs. Through the languages and texts of the voucher policies, particular norms are produced to become regimes of truth. In other words, while promoting "freedom to choose" rules and norms of how to choose and a new definition of the "modern" self are inscribed through voucher policies.

For example, as vouchers are for five-year-old children in licensed private kindergarten and child care programs, when parents are exercising their "freedom to choose," they need to be aware of which programs have (or do not have) government licenses. Not only are the parents being disciplined by the rules of voucher policies, but the field of early childhood education and care is also being regulated through the licensing-granting process. In other words, through voucher policies, parents are simultaneously being governed, and governing themselves, as their choices are shaped by the rules of voucher policies to think of licensed programs as "appropriate" high "quality" or "normal" early educational and child care institutions (Dahlberg, 2000; Dahlberg *et al.*, 1999).

Conceptualizing early childhood educational voucher policies as a sociocultural governing technology that regulates, normalizes, and administers the parents, as well as the field of early childhood education, will lead to a reconceptualization of the concept of social inclusion/exclusion (Popkewtiz, 2000a). As a sociocultural governing technology, educational voucher policies (re)define and produce new norms of "good" and "appropriate/ eligible" kindergarten and child care programs through government licensing and voucher granting processes. Kindergarten and child care programs without licenses thus become "abnormal" or "inappropriate" cultural institutions for young children. Parents who enroll their children in nonlicensed programs are not only being excluded from being rewarded with educational vouchers for their children. They are also being included, or perceived, in a category of "abnormal/bad" parents. As much as the political rhetoric of voucher policies publicizes notions of social democracy and greater social inclusion by framing vouchers as being for all five-year-old children, social exclusion occurs simultaneously.

Conclusion

Contemporary early childhood educational policies have not only interjected new ways of reasoning about what "modernized" education means. They have also introduced a notion of choice into what it means to be "modern" in the field of early childhood education and care in Taiwan. Despite common/popular perceptions of early childhood educational vouchers as a liberal practice and a form of "social justice" for parents with children in private programs, through the different sections of this chapter

I have asserted that it is dangerous to think of voucher and choice discourses/policies as universal magical, progressive, and modern democratic reform practices. My efforts to deconstruct the mentalities which sustain vouchers and school choice discourses, and the "regime of truth" that is produced through voucher policies are motivated by a Derridian notion of ethical attitude (Derrida, 1988).

Little attention has been given to discussions of how voucher policies produce illusions of social democracy in Taiwan. Instead, as voucher and choice discourses have become a new truth in Taiwan, public debates on the topic of vouchers have shifted to discussions on the amount of money allocated to each family.[15] Such popular discussions on issues concerning the face value of educational vouchers underscore a lack of comprehensive understanding of the effects of educational voucher policies, as constructions of new norms are taken for granted without critical reflection. Therefore, out of concern for ethics, my main concerns and arguments within this chapter are to highlight how voucher policies should be thought of as socio-cultural governing technologies to fabricate new ways of reasoning/thinking about what "modern" education means as new "norms" are produced. While today's voucher policies mobilize ideas of choice in Taiwan, my analysis points out that early childhood educational voucher policies are filled with illusions of social democracy.

Lather's utilization of a Foucauldian framework for analysis of educational policies dispels the illusion of vouchers as guaranteeing social democracy. Her work captures the dangers of voucher policies.

Policy is to regulate behavior and render populations productive via a "biopolitics" that entails state intervention in and regulation of the everyday lives of citizens in a "liberal" enough manner to minimize resistance and maximize wealth stimulation. Naming, classifying, and analyzing, all work toward disciplining through normalizing. Such governmentality is "as much about what we do to ourselves as what is done to us" (Lather, 2004, pp. 23–24).

To conclude, if we are not aware of how policies redefine who we are and rework new norms to uphold as "truth," we risk supporting a restrictive and narrow perception of a normative discourse (Cannella, 1997).

Notes

1. For international examples of voucher policies and choice discourses, see the edited work by Plank and Sykes (2003).
2. Highlighting the 1940s as a historical marker is a reflection on the political/regime changes/conflicts between the Republic of China (R.O.C.) and the People's Republic of China (P.R.C). Kuomingtang (KMT) is a political party that was first founded by Dr. Sun Yat-Sen, the Father of Republic of China. After 1949, the KMT moved to Taiwan to carry on the name of the Republic of China while the Communist Party in mainland China took on the name of the People's Republic of China. This chapter discusses Early Childhood Educational Voucher Policies in Taiwan since the late 1990s.

3. The field of early childhood education and care is not been part of the national standardized compulsory education system in Taiwan.

4. Prior to the first enactment of the Preschool Education Act in 1981, the term "Kindergarten" was used to describe early childhood programs before the 9 years of compulsory education in Taiwan. However, after the first enactment of this Policy in 1981, the term "Kindergarten" has only been used to designate programs with educational emphases that are governed and regulated by the Ministry of Education. The original policy texts of the Preschool Education Act from 1981 can be located from a website at http://law.moj.gov.tw

5. In June of 2003, the Preschool Education Act was modified by adding detailed sublines to the original 25 points.

6. The translated and abstracted text of the Preschool Education Act of 1981 from http://law.moj.gov.tw

7. Examples of development discourses and models of early childhood education and care programs from the "well-developed" Western nations can be seen in the multiple Country Notes that are published by Organisation for Economic Co-operation and Development (OECD). Another example is a book entitled *Starting strong: Early childhood education and care*, which provides a comparative analysis of major policy developments and issues in 12 OECD countries, highlighting innovative approaches and proposing policy options that can be adapted to varied country contexts. *Starting Strong* is an e-book by OECD available at http://www1.oecd.org/publications/e-book/9101011E.PDF.

8. The official website for the Association of Early Childhood Education in Taiwan is http://www.eceac.org.tw.

9. Children from low-income families or children of disabled parents can be enrolled in public kindergartens or childcare programs without winning a lottery game. However, the official poverty line is often criticized, as the definition of "low income" is somewhat blurry in Taiwan.

10. The average yearly tuition for private kindergartens during the 1999 school year was about NT $100,000, which is equal to US $2859. The average yearly tuition for public kindergartens during the 1999 school year was about NT $34,600, which is equal to US $989 (*China Daily Newspaper*, March 27, 1999).

11. Mr. Shui-bian Chen was reelected as President in Taiwan for his second term in 2004.

12. NT $10,000 dollars is equal to about US $286. The average income in 2001 of Taiwanese people was about US $14,000.

13. After the first voucher policy in Taipei, many different cities and counties followed these policies that distributed different amounts of money to parents. For example, in the school year of 1998–1999, in Kaohsiung, the voucher was NT $5,000 dollars per school year (this amount of money equals US $143).

14. See note 12.

15. Current proposals to "fine tune" early childhood educational vouchers are extending the amount of vouchers. Instead of granting vouchers in the amount of NT $10,000, new proposals are raising the amount to NT $40,000. Furthermore, instead of giving vouchers for five-year-old children in private programs, new proposals are considering granting vouchers to children in public programs but in lower amounts.

References

Apple, M. W. (2000). *Official knowledge: Democratic education in a conservative age* (2nd ed.). New York: Routledge.

Apple, M. W. (2001). *Educating the "Right" way: Markets, standards, god, and Inequality*. New York: Taylor and Francis.

Bloch, M. N., & Blessing, B. (2000). Restructuring the state in Eastern Europe: Women, child care, and early education. In T. S. Popkewitz (Ed.). *Educational knowledge: Changing relationships between the state, civil society, and the educational community* (pp. 59–82). Albany: State University of New York Press.

Bloch, M. N. (2003). Global/local analyses of the construction of "family-child welfare." In M. N. Bloch, K. Holmlund, I. Moqvist, & T. S. Popkewitz (Eds.), *Governing children, families, and education: Restructuring the welfare state* (pp. 195–230). New York: Palgrave.

Bourdieu, P. (1984). *Distinction: A social critique of the judgment of taste.* Cambridge, MA: Harvard University Press.

Cannella, G. S. (1997). *Deconstructing early childhood education: Social justice & revolution.* New York: Peter Lang.

Carnoy, M., & McEwan, P. J. (2003). Does privatization improve education? The case of Chile's national voucher plan. In D. N. Plank, & G. Sykes (Eds.), *Choosing choice: School choice in international perspective* (pp. 24–44). New York: Teachers College Press.

Cookson, P. W. (1994). *School choice: The struggle for the soul of American education.* New Haven, CT: Yale University Press.

Dahlberg, G. (2000). From the "People's Home"—Folkhemmet—to the enterprise: Reflections on the constitution and reconstitution of the field of early childhood pedagogy in Sweden. In T. S. Popkewitz (Ed.), *Educational knowledge: Changing relationships between the state, civil society, and the educational community* (pp. 173–200). Albany: State University of New York Press.

Dahlberg, G., Moss, P., & Pence, A. (1999). *Beyond quality in early childhood education.* London: Routledge Press.

Derrida, J. (1988). Letter to a Japanese Friend. In D. Wood, & R. Bernasconi (Eds.) *Derrida and Difference.* (pp. 1–5). Evanston, IL: Northwestern University Press.

Escobar, A. (1995). *Encountering development: The making and unmaking of the Third World.* Princeton, NJ: Princeton University Press.

Foucault, M. (1991). Governmentality. In G. Burchell, C. Gordon & P. Miller (Eds.), *The Foucault effect: Studies in governmentality* (pp. 87–104). Hemel Hempstead: Harvester Wheatsheaf.

Foucault, M. (1980). Truth and power. In C. Gordon (Ed.), *Power/Knowledge: Selected interviews & other writings, 1972–1977* (pp. 109–133). New York: Pantheon Books.

Gewirtz, S., Ball, S. J., & Bowe, R. (1995). *Markets, choice and equity in education.* Philadelphia, PA: Open University Press.

Lather, P. (2004). This IS your Father's paradigm: Government intrusion and the case of qualitative research in education. *Qualitative Inquiry 10* (1), 15–34.

Lu, M. K. (1984). *Early childhood educational policies.* Taipei, Taiwan: Wen-Ching.

Lu, M. K. (2001). An evaluation and analysis of the early childhood educational voucher policy in Taipei City. *The Journal of Municipal Teacher's College of Taipei City, 32,* 22.

Lu, M. K., & Hsieh, M. H. (2001). *Early Childhood Educational Voucher: Theory and Practice.* Taipei, Taiwan: Teacher's College Press.

Lu, M. K., & Shih, H. Y. (2000). The feasibility of the early childhood educational voucher. *The Journal of Municipal Teacher's College of Taipei City,* 31, 161–192.

Nóvoa, A. (2002). Ways of thinking about education in Europe. In N. Novoa & M. Lawn (Eds.), *Fabricating Europe: The Formation of an Educational Space* (pp. 131–156). Dordrecht: Kluwer Pubishers.

Pan, J. L. (2000). "1018 Walk for Early Childhood Education: Vouchers are a form of social justice." Retrieved April 12, 2002, from http://www.yuanlin.gov.tw/MAGZINE/PAGE/yl11_03.htm

Plank, D. N., & Sykes, G. (2003). Why school choice? In D. N. Plank, & G. Sykes (Eds.). *Choosing choice: school choice in international perspective* (vii–xxii). New York: Teachers College Press.

Popkewitz, T. S. (2000a). National imaginaries, the indigenous foreigner, and power: comparative educational research. In J. Schriewer (Ed.), *Discourse formation in comparative education* (pp. 261–294). Berlin: Peter Lang Verlag.

Popkewitz, T. S. (2000b). Rethinking decentralization and the state/civil society distinctions: the state as a problematic of governing. In T. S. Popkewitz (Ed.), *Educational knowledge: Changing relationships between the state, civil society, and the educational community* (pp. 173–199). Albany: State University of New York Press.

Popkewitz, T. S., Pereyra, M. A., & Franklin, B. M. (2001). History, the problem of knowledge, and the new cultural history of schooling. In T. S. Popkewitz, B. M. Franklin, & M. A. Pereyra (Eds.), *Cultural history and education: Critical essays on knowledge and schooling* (pp. 3–44). New York: Routledge Falmer.

Popkewitz, T. S. (2003). Partnerships: The social pacts and changing systems of reason in a comparative perspective. In T. S. Popkewitz, B. Franklin & M. Bloch (Eds.), *Educational Partnerships and the*

State: The Paradoxes of Governing Schools, Children, and Families. (pp. 27–54). New York: Palgrave Macmillan.

Rose, N. (1998). *Inventing our selves: Psychology, power, and personhood.* Cambridge: Cambridge University Press.

Rose, N. (1999). *Powers of freedom: Reframing political thought.* Cambridge: Cambridge University Press.

Shore, C., & Wright, S. (1997). Policy: A new field of anthropology. In C. Shore & S. Wright (Eds.), *Anthropology of Policy* (pp. 3–42). New York: Routledge.

Taipei City Government (1997). Retrieved April 12, 2002, from http://www.look.taipei.gov.tw/

CHAPTER 11

The Foundation Stage Child in a Shifting Sea: A History of the Present of the United Kingdom's Education Act 2002

RUTH L. PEACH*

Many chapters in this book deal with circulating global or international discourses about young children; this chapter is one of them.[1] Focusing on the inclusion of young children in an educational policy, the Education Act 2002 from the United Kingdom, I look at circulating international discourses within it, while highlighting the ways in which local factors such as politics, culture, and history interplay with the effects of power that created, and are created by, this policy. I argue that the substantive developments in the field legislated by the Education Act 2002 are problematic and may propagate the same inequities that they are intended to solve. This is the case for educational reform in multiple contexts; the U.K. policy is the example I use in this chapter.

This piece of legislation, written in the early years of the twenty-first century, lies among and within historical discourses about the young child, the good citizen, schooling, teaching, and parenting. As it emerges, an apparently new and discrete entity, those past trajectories push and pull at the child at the same time as they contribute to its shaping. While this child is being shaped by historical discourses, the discourses also shift as new ways of knowing the child emerge: "An 'age' does not pre-exist the statements which express it, nor the visibilities which fill it" (Deleuze, 1986/1988, p. 48). The U.K. policy articulates a shift in the "age" of young citizens, creating and created by bringing the child further into the public sphere than it had been. In this chapter, I historicize this shifting child and distinguish some of the invisible visibilities within, around, and before the Education Act 2002.

This chapter looks at multiple traveling discourses within this Act. The first is the movement from the human beings seen as an economic resource, the Third way and neo liberalism to the more recent marketization of education. The second is the binary of the "at risk"/normal child and universalized scientific knowability. The third is the concept of a panopticon, with the idea of surveillance and control exercised at a distance.

Theoretical Framing

This chapter is framed using Foucault's ideas of truth and power. Foucault posits that truth and power are inextricably intertwined. We perceive and create reality through certain types of culturally and historically situated beliefs or "truths."

> Truth is a thing of this world: it is produced only by virtue of multiple forms of constraint. And it produces regular effects of power. Each society has its régime of truth, its "general politics" of truth: that is the type of discourse which it accepts and makes function as true: the mechanisms and instances which enable one to distinguish true and false statements, the means by which each is sanctioned; the techniques and procedures accorded value in the acquisition of truth; the status of those charged with saying what counts as true. (Gordon, 1980, p. 131)

Truth is not universal but is the product of a particular time and place. Through our understandings of what is true and what is not true, veins of power run through our society.

Truth is also not historically constant, but is marked by disjunctures where one understanding of truth is replaced by another. In this chapter, I look at how the Education Act 2002 is historically and nationally situated. I explore the history or genealogy of the discourses or the underlying "truths" that make up the "reality" of the young child that this policy creates and is created from.

In the first section I look at a few of the ways the child appears in the policy, historicizing shifts produced by this policy in the sea of early years education and care in the United Kingdom: the young child as foundational within the trajectory of the educated young U.K. citizen and the young child as scientifically knowable and governable through the application of standards, high-stakes tests, and national curriculum. The second section discusses some of the ways the policy moves *in* the child, through global trajectories within and around the Education Act 2002: the situation of the young child as a young citizen who is part of the nation/state's human capital, and the binary of the at risk/normal child.

The Shifting Young U.K. Citizen: The Child in the Policy

How are young children actually included in the text of the Education Act 2002? Before asking that we first must ask, can a child be defined by a policy?

> The familiar child in a comfortable "box" of recognition in the present
> is therefore troubled by the historical work that catalyzes that recogni-
> tion toward a loss of Familiarity . . . What one might know or think
> about the child at the outset, the consoling play of recognitions, is there-
> fore problematized by "moves" across discursive space, securing at the
> end an ambiguity, uncertainty, and strangeness. (Baker, 2001, p. 52)

This ambiguity, uncertainty, and strangeness trouble the fast-moving, deci-
sive, problem-and-solution-oriented waters of policy language. In order for
a child to be the location of reform, not only must that child be at risk, not
the norm, and in need of salvation. That child must also be knowable,
defined, and bounded. Further complications of "truths" about the child
point to recent shifts in the imaginary of the child within the national imaginaries of young children which occurred in part because of research in the
area of brain research in infants, in part because of new policies and legislation providing early years learning which have arisen in the global discourse
about young children. As brain research in preverbal infants (and in utero
fetuses) showed even very young children to be active individuals and
learners, beings with potential which can be enhanced or expanded
through the application of the correct methods, the caregiver(s) of young
children became the site of increased direction from psychologists and legislators, as specific practices were prescribed to ensure a normal (or supernormal) child. This research played a role in reshaping the child who was
to be acted upon by the early years education policy. So within this ambiguity, how is the child framed in the Education Act 2002?

The Education Act 2002 created a new category of child, the "Foundation
Stage child," located before the beginning of formal publicly funded schooling. This Act distinguished three- to five-year-old children from infants,
introduced a specific national curriculum for the Foundation Stage child, and
placed these new children within the Ministry of Education instead of the
Ministry of Health that had previously been responsible for the age group.
The policy defines this new Foundation Stage child, approximately 3–5 years
of age (Education Act 2002 [c. 32], 2002, p. 56), in relationship to the Key
Stages One through Four, a previously established system covering primary
and secondary schooling; Key Stage One starts at the beginning of primary
school, and Key Stage Four ends when the child is no longer required to
attend school (approximately age 16). Defining the three- to five-year-old
child as part of the Key Stage system situates the three- to five-year-old child
at the beginning of a linear progression from "school child" to "adult,"
which is a major shift from previous education policies in England. This new
positioning of young children locates them as part of the human capital of the
country, which I will talk more about in a later section.

The Labour Party passed their promised reform, the Education Act
2002. The Education Act 2002 home page on the Department for

Education and Skills website (http://www.dfes.gov.U.K./educationact 2002/) describes the role the Act is to play: "It is a substantial and important piece of legislation intended to raise standards, promote innovation in schools and reform education law." Children under the age of primary school had not been included in earlier Education Acts. Education Act 2002 was preceded by Education Acts every 1–5 years starting in 1944, but earlier versions do not include sections that change the programs for early years children to such a degree. In contrast, the U.K. Education Act 2002 states very early in the document that it applies to "persons under the age of nineteen (Education Act 2002 [c. 32], 2002, p. 2) and that it is "an Act to make provision about education, training and *childcare* [italics added]" (Education Act 2002 [c. 32], 2002, p. 1). This legislative definition of the child in Education Act 2002 includes all young children in the United Kingdom, without a definite lower age limit. The Education Act 1998, the Act immediately prior to the 2002 Act, defines the children being served under the Act as primary and postprimary students only. Where were young children located in the legislative galaxy before they were "discovered" and moved into the forefront in the Education Act 2002? They were tucked away in the Child Care Act 1991, which has sections regulating children's health, care, welfare, residential centers, adoption, prevention of cruelty to children, and child care.

Moving the three- to five-year-old child from jurisdiction of the Ministry of Health to the Ministry of Education implies that in the education care spectrum of early years programs this move would increase the emphasis on education and define young children in relationship to schooling rather than infancy. As I mentioned earlier, shifting young children into the realm of schooling also implies moving them into the arena of human capital. Young children are now a resource to be cultivated, with hopes of future financial gain if that resource is developed wisely, but also fear of potential future loss if the resource is squandered. Consequently, young children, as a potential economic resource, are now subject to surveillance. Young children now have a defined curriculum that is a matter of national interest, and the programs are subject to state inspections.

This shifting definition of young children as located within schooling rather than health is certainly not unique to the United Kingdom, and it remains to be seen how it will impact young children. Young children were not included in the National Curriculum until the Education Act 2002 established the Foundation Stage curriculum. Early childhood professionals have mixed positions on the shift, mostly agreeing that both education and care are important but, while supporting the increasing recognition and professionalization of the field, many have important concerns about pressuring young children into too much academic programming too young. One critique of this new form of "curriculum shovedown" warns:

My view is that the standards movement that seems pervasive across educational settings is threatening children in early childhood in the same ways as the curriculum shovedown movement of the 80's. The point of attack has changed from curriculum to outcomes, but the consequences for young children may be the same. (Hatch, 2002, p. 457)

This critique of the standards and accountability movement into early years programs is timely and important in light of the directions that this reform is taking early childhood education in.

A significant portion of this Act in relation to early years learning is "Financial Assistance for Education and Childcare," which provides funding for Foundation Stage children for 2 1/2 hours per day, five days per week (Education Act 2002 [c. 32], 2002, p. 10), paid directly to the program the child attends. In addition, funds are available to start up new Foundation Stage programs as well as to add Foundation Stage to existing early years and primary programs. The child in the Education Act 2002 is now a child who is foundational to the years of schooling, guided by a national curriculum, and deemed worthy of public funding. The next section situates this new Foundation Stage child within the policy that created it and situates that policy within the sea from which it arose in the United Kingdom in 2002.

Global Trajectories in Education Act 2002:
The Policy in the Child

The sea of globalized trajectories or traveling discourses within and surrounding the production of the Education Act 2002 created, and were created by, what was thinkable and knowable at their historico-social location at the beginning of the twenty-first century. These traveling discourses circulate through the education policies and the practices of multiple nations: the first of these discourses that I will discuss in this section are the marketization of the child that shifts schooling to a business model and locates the young child as human capital; the second is the trajectory of normalization and the at risk/normal binary; the third is the technologies of surveillance that are included in the Act and its formulation. Before discussing these three traveling discourses as they appear in the Education Act 2002, I offer a brief genealogy of neoliberalism and the Third Way. Part of the shift that moved schooling to a business model and the young child into the arena of human capital for the nation/state, both neoliberalism and Third Way were the sea from which marketization of schooling emerged.

Governmentality and Populational Resources

The idea of a "Third Way" did not come out of a historical vacuum. In the following part of this section I look at the historically situated circumstances that made the emergence of these ideas possible and at the specific and historically situated shifts in understanding that have made market-oriented framings of the young child possible and even natural. Michel Foucault (1988) in his essay, "Governmentality," describes the first of these shifts. Governmentality refers to a shift that occurred during the eighteenth century (cited in Martin *et al.*, 1988, p. 103) in which the locus of control moved from a circulation of power between the sovereign and the individual to a circulation of power in which individual citizens regulate themselves. Governmentality moves the regulation of individuals to ways that are "at once internal and external to the state" (p. 103) as they strive to be "good" citizens. This production of power is one in which "[r]elations of power are not in a position of exteriority with respect to other types of relationships . . . they have a directly productive role, wherever they come into play" (Foucault, 1976/1990, p. 94). The internalized weaving of the web of power relationships is reflected particularly strongly in the part of this education policy regulating governmental inspections of Foundation Stage programs, which I discuss further later in the "Surveillance" section of this chapter.

In governmentality Foucault says that the concept of wealth changed from the control of territory, as was true of feudal lords in the Middle Ages, to the control and disposition of human resources. A nation's wealth is no longer entirely in its geographical resources, but also in the skills, talents, and attitudes of its inhabitants who now are citizens, not serfs. This shift led to the development of sophisticated statistical apparatuses for understanding and controlling data about the population. The late eighteenth and early nineteenth centuries saw an explosion of interest in census data and in statistically based social sciences, such as sociology. Because of this interest the U.S. census, for example, has expanded from the original four questions to hundreds of questions in the long forms of modern census.

A concern for the development of populational resources also made it seem natural to put public funds into universal schooling. The nineteenth and early twentieth centuries saw public schooling go from a new idea to an unquestioned public responsibility in multiple nations, including the United Kingdom. By the mid-nineteenth century, an educated populace was synonymous with civilization.

Human Capital, Neo-Liberalism, and the Third Way

The idea of controlling the population was given further scientific framing during the middle years of the twentieth century with the prominence given to "human engineering" or the "rational utilization of the human factor in the management of institutions and society" (Rose, 1999b, p. 15) during World War II. Psychology, sociology, anthropology, and industrial management all became strongly preoccupied with understanding and classifying the members of the population and workforce. The purpose of these understandings was to develop human resources for a rational and efficient military and civilian workforce.

The 1950s saw an international increase in concern for developing children as a future labor pool. By the late 1950s and early 1960s, this concern had formalized into human capital theory. Human capital has been very influential in the business community since Jacob Mincer published his groundbreaking article, "Investment in Human Capital and Personal Income Distribution" in *The Journal of Political Economy* in 1958. The idea was further developed by Gary Becker of the Chicago School in his book, *Human Capital,* originally published in 1964:

> To most people capital means a bank account, a hundred shares of IBM stock, assembly lines, or steel plants in the Chicago area . . . economists regard expenditures on education, training, medical care, and so on as investments in *human* capital (Becker, 1993, ¶ 1).

Harbison and Meyers (1964) describe the urgency embedded in human capital. "Advanced" countries have highly developed human resources; countries that do not properly invest in and control their human resources become or remain underdeveloped (Lightfoot, 2001).

> In an advanced economy, the capacities of man are extensively developed; in a primitive country they are for the most part undeveloped. If a country is unable to develop its human resources, it cannot develop much else, whether it be a modern political structure, a sense of national unity, or higher standards of material welfare. Countries are underdeveloped because most of their people are underdeveloped. (Harbison & Meyers, 1964, p. 13)

In the 1980s, this concern about human capital and the development of the nation's resources created a new idea, neoliberalism. Nikolas Rose and others locate a move to a market focus within neoliberalism: "The theme of enterprise that is at the heart of neoliberalism certainly has an economic reference" (Rose, 1999a, p. 230), a shift from the welfare state and human capital model which preceded neoliberalism.

More recently, in the 1990s, the discourse of the Third Way emerged. Anthony Giddens (2003) popularized the term and authored multiple publications on the Third Way. He defined the Third Way in terms of what it replaced:

> The third way is *not* a "middle way"—specifically, it is not an attempt to find a halfway point between the Old Left and free market fundamentalism. It seeks to transcend both of these. Neither of these earlier two "ways" is adequate to cope with the social and economic problems we face today (p. 2).

The relationship between Third Way and neoliberalism is debated even by those who see them as part of the same political/economic theory; one side of the debate says that "the so-called third way is simply neoliberalism with a human face" (Isaac, 2001, p. 1) while Anthony Giddens who is the Third Way's most prominent mouthpiece answers these critics: "Third Way politics is not a continuation of neoliberalism, but an alternative political philosophy" (Giddens, 2000, p. 32). This is a debate with interesting facets; for the purposes of this chapter I assume that neoliberalism and the Third Way are part of the same complex of trajectories that produced, and were produced by, the shift to a market-based political model in England and in many other nations, and that they were a prominent part of the sea which produced the Education Act 2002.

"Third Way was developed above all as a critique of the neoliberal right" (Giddens, 2003, p. 6) when it propelled England's Prime Minister Tony Blair into office after years of conservative government in the United Kingdom, uttering his now-famous promise to focus on "education, education, education" (Curtis, 2003). After a weekend with former U.S. president Bill Clinton in 2002, referred to as a "third way thinkathon" (Wegg-Prosser, 2002, p. 1), Tony Blair renewed the commitment to education he made six years earlier in a campaign promise. Situated within and as an integral part of the Third Way and neoliberalism, the move to use a business model to reform education became a globalized circulating discourse that many countries adopted. Human capital theory urges support for education because of the belief that "large increases in education and training have accompanied major advances in technological knowledge in all countries that have achieved significant economic growth" (Becker, 1993, ¶18). The U.K. Education Act 2002, along with the U. S. *No Child Left Behind* Act of 2001, written and passed within months of the Third Way thinkathon, are two reforms that legislated a business model or marketization of education for young children.

Marketization of Education

Moving young children into the pool of human capital for the nation/state highlights their value and importance, a goal for early years educators in many countries, but it also produces young children in new ways that may be problematic. One way that the shifting production of young children as part of the human capital of the nation/state through these reforms is that they are now included in the national policy governing publicly funded schooling. The Education Act 2002 moved early years children (ages 3–5) from the jurisdiction of the Ministry of Health to the Ministry of Education. There are benefits for early childhood professionals in this new production of young children as human capital such as increased professional respect, but at the same time this shift changes the production of young children and their schooling. Part of this shift is that early years children, staff, and curricula become the locations of the practices of school inspections, national curriculum, and high-stakes testing that were implemented in state-funded schooling in the 1990s. These practices were part of the Third Way move to marketization of education that promised teacher accountability and measurable results on high-stakes tests, among other results. Marketization moves the trajectory of education so that it will produce

> a strong public sphere, coupled to a thriving market economy; a plu-ralistic, but inclusive society; and a cosmopolitan wider world, founded upon principles of international law. (Giddens, 2003, pp. 6–7)

Producing young children as human capital for their countries through a business model of schooling was part of the political agenda behind the Education Act 2002 in the United Kingdom and other reforms in other countries.

Politicians promoted these reforms by promising that their enactment would create improvements in the quality and quantity of workers. The fol-lowing excerpt from a speech promoting the U.K. reform is one example:

> The next few years pose a special challenge—to move from catching up with the rest of the world, as we have by cutting class sizes, raising teacher salaries, improving pedagogy, to moving ahead and giving our young people the best possible chance of making their way in the world and contributing to *economic and social renewal* [italics added] in this country. (Miliband, 2002)

In another example, Ms. Estelle Morris, former Minister for School Standards, included this comment in a report made to Parliament on December 16, 1999, as the Labour party was in the process of creating the education program that produced the Education Act 2002:

> The Government have established child care as a major strand of their school standards agenda, of their family-friendly policy and of *economic and competitive policy* [italics added]. (United Kingdom Parliament, December 16, 1999.)

This emphasis on creating a U.K. citizen who will be competitive in a globalized economy demonstrates the human capital rationale behind increased support for early years programs.

A third example comes from the United States. At around the same time as the production of the Education Act 2002 in the United Kingdom, the *No Child Left Behind* Act of 2001 was passed in the United States. The content of this quote from a speech by the U.S. Secretary of Education contains lan-guage that is quite similar to that used by the U.K. officials in the speeches cited earlier; this similarity indicates a circulation of ideas between the two countries. When signing the *No Child Left Behind* Act in 2002, the U.S.

Secretary of Education Rod Paige under President George W. Bush said:

> Instead of paying for services, we will be *investing in achievement* [italics added]. When federal spending is an investment, it gives the federal government leverage to demand results. And demanding results is what the Department of Education will do. (Paige, 2002)

The content of the three comments makes the business model/human capital educational goals in use at the time these two reforms were written quite transparent. Investing in achievement and demanding results from that investment moves teachers, children, parents, and schooling into a marketized relationship. Marketization, high-stakes testing, standards, and schooling as investment combine to redefine schooling for young children in the Education Act 2002 reform, as in other reforms in various nation/states.

Many studies and analyses of the marketization of schooling implemented in state-funded U.K. schools in the early 1990s were done; several of these were highly critical (e.g., Mahoney & Hextall, 1997; Salisbury & Riddell, 2000). One, a 1992 study by Bowe & Ball, noted that

> The ideology and political rhetoric of the market, as directed towards the welfare state, celebrates the superiority of commercial planning and commercial purposes and forms of organization against those of public service and social welfare (p. 53).

The tensions between policymakers who viewed state-funded schooling as a vehicle for producing skilled workers or well-rounded human beings are not new, having been present when universal compulsory education started (see Ball, 1990), and while these are not discrete educational goals, the balance has shifted under differing governments and changing social goals. Echoes of this tension appear in the current shift to extend marketization to younger citizens.

A recent U.K. study surveyed early years practitioners about their experiences of the marketization shift in their programs:

> The overwhelming majority of practitioners felt business-like approaches to management were inappropriate in childcare and instead emphasized the importance of collaboration and mutual support as their preferred way of operating. (Osgood, 2004, p. 10)

Along with these studies, the "Accountability Shovedown" critique by Hatch (2002) cited earlier bring up several ways that the intersection between business models of schooling and early years programs may be problematic. Due to the fairly recent implementation of this reform, long-term studies of the business model on early years children and programs are not yet available. The earlier use of the business model in primary and secondary schools has generated some changes in the way the model is carried as a result of critiques; it is too soon to know what this shift might bring to the younger children and their caregivers and families.

At Risk/Normal Binary

We saw above that the concept of human capital is based on a binary—that of "developed" or "underdeveloped" human resources. The implicit threat is that some of its citizens will fail to develop properly. In this way, they are at risk and through their personal failure they put the nation at risk.

Multiple reform policies have been created with the goal of saving the "at risk" child which has become code for uneven access to resources and a hierarchialization of knowledge. The Education Act 2002 is one of these; in a speech by Education Secretary Morris in the month the Act was signed, she said, "we need to focus attention, in far greater depth then we have had so far, on the concentration of disadvantaged pupils and under-performing schools" (Morris, 2002). I have pulled out three circulating discourses related to risk embedded within the language of this policy that particularly shapes it. The first is the at risk/normal binary; the second links nonnormality with pathology of the social body; and the third is the application of standardized salvation through educational reform.

The term "at risk" contains an embedded other, the "normal" child, which defines the at risk child by its illuminating what it is not and, by definition, will never be (e.g., Dahlberg *et al.*, 1999; Swadener & Lubeck, 1995; Tabachnick & Bloch, 1995). Rather than being created to save the at risk child, reform policies *require* that some children be designated as at risk in order to have a location for the social actions on the part of the reformers and to isolate the at risk child from the normal. The child designated as at risk and who is the site on which education reforms are to act tends to be a child who is marginalized by a difference, which is not the norm, by definition. Therefore sequential reforms continue like a cat chasing its tail and can never succeed or be complete within the binary and oppositional language of normal/other.

> The scientific and normal child is a discourse that arose around the end of the nineteenth century and early part of the twentieth. Throughout the nineteenth century, a combination of science laced with romanticism and enlightenment beliefs resulted in new discourses in which religious ideas of salvation after death shifted toward progress, modernization, and ameliorating life "on earth" through science and reason. (Bloch *et al.*, 2003, pp. 15–17)

A child who does not perform according to the projected scientific criteria is labeled at risk, therefore abnormal, ungovernable, and marginalized.

Social inequities and cultural differences are ignored in this discourse of at risk, with standards and high-stakes testing assuming a universal and scientifically knowable child who, upon application of the correct curriculum, will produce certain answers. The emphasis upon universal standards, including high-stakes test results, in the new reforms perpetuates the myth that the scientific transcends cultural and social differences. These supposedly universal scientifically derived standards that are used to create the normal/at risk binaries have been frequently critiqued (see Swadener & Lubeck, 1995; Tabachnick & Bloch, 1995); in the very production of developmental standards, some knowledge(s) and types of thinking are highly privileged over others, though this is often not taken into consideration.

> Scientific research was a critical strategy used to construct truth about who was normal and which children or families were perceived as abnormal and in need of different social interventions. Statistical categories for the normal family were constructed from narrow samples and contrasted with demographic population facts about different families and different cultural/social/religious/economic organizations. (Bloch, 2003, p. 206)

The universal skills tests actually measure how closely a child resembles the normal, which is a gendered, racialized, classed category even though it is regarded as value-free and universal.

The use of statistics and populational reasoning is one way that the normal child came into being:

> The normal was one of a pair. Its opposite was the *pathological* [italics added] and for a short time its domain was chiefly medical. Then it moved into the sphere of—almost everything. (Hacking, 1990, pp. 160–161)

Linking normality with pathology, as Hacking does, leads to a second circulating discourse within this policy. Not only was nonnormality linked with disease, it moved into a prediction about the future of the child who was at risk. The designation of diseased/at risk loses the rich depth of experience and strength of diverse lives into the flattened categorization of not normal, suspect, or diseased.

> Disease is perceived fundamentally in a space of projection without depth, of coincidence without development. There is only one plane and one moment (Foucault, 1963/1973, p. 6).

The child who is defined as diseased, without depth, without development, is not an actual child but is one who is created in the minds and hearts of reformers, to infuse them with zeal for the reform and their efforts. What is the response of society to the diseased child?

> Let us call tertiary spatialization all the gestures by which, in a given
> society, a disease is circumscribed, medically invested, isolated,
> divided up into closed, privileged regions . . . it brings into play a sys-
> tem of options that reveals the way in which a group, in order to pro-
> tect itself, practises exclusions, establishes forms of assistance, and
> reacts to poverty and to the fear of death. (Foucault, 1963/1973,
> p. 16)

The figure of the nonnormal child is separated in order to protect the
norm, to locate poverty at a distance with the "other," along with death.

A third discourse of normalization found in the Education Act 2002 and
many others describes the normal child through standards, the National
Curriculum, and high-stakes testing. Standards and national curricula for
young children are based on the scientific gaze in which children were seen
as universally knowable, following normed steps and stages of development. The tension between the cultural focus of multicultural education
and the universal, scientific norms of child development became heightened with some legislators, caregivers, and families shifting from diversity
education to concerns with standards, assuming a universal, scientifically
knowable child. The scientific and normal child is a discourse that arose
around the end of the nineteenth century and early part of the twentieth.
This shift has been described as follows:

> The new knowledge from nineteenth-century missionary reports,
> travelers' reports, and anthropologists' and economists' reports were
> used to draw conclusions about "civilization" and "culture" and the
> "nature" of primitive families and childhood in exotic places (see Said,
> 1978). These reports, particularly when based on scientific observa-
> tion of others, were used to draw conclusions about universal laws for
> developing children to become civilized. (Bloch *et al.*, 2003,
> pp. 15–17)

In the sea of national education policy in the United Kingdom, young children have not been welcomed in state-financed schooling but have followed in its large wake in a motley collection of small and modest craft,
most powered solely by family-paid tuition costs. These small craft vary
from tiny gems of luxury with all the frills to leaky barrels cobbled together
of bits of this and that, barely afloat. This very unequal and uncertain fleet
has been critiqued in the OECD report (Bertram & Pascal, 1999) and in
many research projects (e.g., Bruner, 1980; Moss & Penn, 1996). In many
countries, including the United States at this time, this is still the case. In
the United Kingdom, the Education Act 2002 has taken a step to change
this, partially through the ways in which it has rewritten the child.

Surveillance

In order to manage the inevitable risk posed by the need to develop a group of individuals as an economic resource, it is crucial to invent and maintain methods of control and surveillance. Foucault envisioned these methods of surveillance through the metaphor of the panopticon, a prison designed in the nineteenth century. This prison design placed prisoners in such a position that they could not see when the guards were present and watching, but would assume that they were always under surveillance and would therefore govern themselves as if they were always under scrutiny.

Control exercised at a distance by the design of the panopticon is reflected in multiple arenas of power/knowledge relationships that are woven into educational discourses, including education reforms. Governing at a distance, or governmentality, was part of the historico-social location, which produced the reform Education Act 2002 and the policies that preceded it along with the practices legislated by those policies.

In the Education Act 2002, the idea of necessary surveillance is manifested through two areas: the way that the Act was formulated, and the required state inspections of early years programs legislated by the Act. The English Education Act 2002 was preceded by the First Report, followed by the Second, Third, and Fourth Reports, which integrated new research as well as critiques from practitioners, input from the public via a website, and expert testimonies into the contents of the First Report. The First Report committee included policymakers, educators, parents and bureaucrats in its formulation, either in initial stages or in subsequent ones. These four reports were completed in 2000, and were used to produce the Education Act 2002. The First Report and white papers were distributed widely, including online, and all readers were encouraged to email feedback to the Committee. The Second, Third, and Fourth Reports were issued as a result of this process. Opening up the Report to multiple participants in early education moved the power/knowledge relationship from that as knowledge held by the experts, sovereign power, to a shared governmentality through which the participants governed themselves.

Educators, who in order to be funded had to follow the National Curriculum and the policies, were then positioned as having been part of the creation of the policies that governed them. The First Report accomplished this, since early years educators were included on the subcommittee that produced the First Report, early years professionals were called upon by the committee to provide feedback to the committee as expert witnesses, and the profession at large was encouraged to provide feedback to the Committee via the website. Information was sent with the free copy of the national curriculum for foundation stage requesting feedback as well. This was intended to create a legislation which was responsive to the practitioners but it also created a situation in which the practitioners "had no one to blame but themselves" if they disagreed with the legislation which arose from the recommendations of the First Report.

The opportunities publicly provided for practitioners to be part of producing the policy placed against the idea of governmentality shows ways in which the policy-formulation process incorporates embedded discourses of self-regulation, no matter what use was made of the actual input and criticism from practitioners.

After the first report was issued, a series of hearings was held in which educators were invited to issue White Papers in response to the First Report and some of these educators spoke before the Select Committee. University experts in early education and care were invited to be advisors to the subcommittee. Their input was followed by oral evidence sessions at which witnesses were able to critique and ask questions about the findings of the committee as well as to provide written White Papers commenting on the contents.

Governmentality also shows up in inspections of early years programs by the Office for Standards in Education (Ofsted), methods by which the Curriculum Guidance for the Foundation Stage as well as all other Key Stages is implemented. Ofsted started in the Education (Schools) Act 1992. This act required Ofsted to regularly inspect all schools that are funded by the nation/state, including nursery and independent schools (Office for Standards in Education, ¶2).

Ofsted inspects voluntary, private, and independent preschools, nurseries, and schools offering the foundation stage curriculum, and the inspector makes a judgment on the quality of education offered by the setting. Anyone can see the Ofsted inspection reports, including those of foundation stage programs, at www.Ofsted.gov.U.K. on the Ministry of Education website or as paper copies, provided free of charge, to anyone who is interested. Each program may be scrutinized by anyone at any time. The program is governed by the Ofsted inspector, and by the public; the inspection reports are posted on a website which is available at any time to the public. The result is that the inspection reports are quite high stake.

What is the format of these reports? Do they allow for the complex, rich, messy portions of early years programs such as learning through play and exploration, or do the reports mould early years teachers to "teach to the test" as some practitioners in English middle childhood programs reported after a national curriculum and school inspections were implemented in their schools (Mahoney & Hextall, 1997)? How does the public surveillance of the documents govern the teachers and administrators, the inspectors? Is there another way in which information about the programs could be made available to the public but that would be more respectful of the practitioners? Is there recourse for practitioners who disagree with the inspection reports? Are all inspectors sufficiently familiar with early years practice to include the subtleties that distinguish a competent program from a great one and to write a report that will communicate that with those reading the report?

The surveillance over the investment that the U.K. government makes in early years programs and in the children in order to shape them into the kind of human capital the nation/state requires in the era of globalization moves early years programs into an important position in the nation's education policy. This movement brings with it changes in the situation of the young child that were not necessarily an intended part of the policy formulation. These changes and the global trajectories situated within the language(s) of the policy raise questions about the many results that are being produced by the Education Act 2002.

Conclusion

Powerful currents are moving the young child all around the globe, floating in the shifting sea of a new identity at this time/space in the early years of the twenty-first century. These currents contain strong eddies, positioning young children at the foundation of state-funded schooling; some eddies are troubling as young children join the marketization of education, becoming human capital for the nation/state in a new way. These problematic currents, however, shift the power/knowledge relationships that young children have had as young citizens, redefining them as worthy of substantial investment, bringing them further into the public arena, and sending needed resources to the programs and caregivers who shape them. High-stakes testing, curriculum shovedown, national curriculum, state inspections— all of these—move the new Foundation Stage child, reforming this age group within the constellation of the educable citizen.

The reforms, because of the binaries of risk/promise contained in their wording, will not save every child from socially constructed hazards and will not provide all the solutions to all of the problems they were written to solve. In spite of this, the reforms have provided, and will continue to provide, a new realization that young children are dynamic participants in the nation/state that nurtures them. Within the global trajectories of early education, in the midst of the swiftly shifting currents, many eyes will watch for the movements created in the United Kingdom by the Education Act 2002.

Notes

* This chapter would not have been possible without the generosity of several colleagues: in the England contingent, thanks to Mrs. Elizabeth Coates of University of Warwick in Coventry for including me in her excellent courses and for her reflections on the changes in English early years policies; to Ms. Gee Fang Lee for sharing her experiences putting the policy into practice; and to Mr. Peter Edworthy for his sharply insightful yet gently offered comments. The responsibility for the use of the information they shared with me is entirely mine. Many thanks to Dr. Theodora Lightfoot, Dar Weyenberg, and Devorah Kennedy for their incisive editing and unrelenting support, and to the members of Thursday Group at the University of Wisconsin-Madison for their inspiration and patience. Finally, many thanks to Professor Marianne Bloch for her generous advising and mentoring.

1. An earlier version of this chapter was presented at the Reconceptualizing Early Childhood Education Conference in Madison, Wisconsin on October 19, 2005.

References

Baker, B. (2001). *In perpetual motion: Theories of power, educational history, and the child.* New York: Peter Lang.

Ball, S. (1990). *Politics and policy making in education: Explorations in policy sociology.* London: Routledge.

Becker, G. (1993). Human Capital. Retrieved September 16, 2005, from http://www.econlib.org/library/Enc/HumanCapital.html

Bertram, A., & Pascal, C. (1999). *The OECD thematic review of early childhood education and care: Background report for the United Kingdom.* Worcester: Centre for Research in Early Childhood.

Bloch, M. (2003). Global/local analyses of the construction of "family-child welfare." In M. Bloch, K. Holmlund, I. Moqvist, & T. S. Popkewitz (Eds.), *Governing children, families & education: Restructuring the welfare state* (pp. 195–230). New York: Palgrave Macmillan.

Bloch, M., Holmlund, K., Moqvist, I., & Popkewitz, T. (2003). Global and local patterns of governing the child, family, their care, and education: An introduction. In M. Bloch, K. Holmlund, I. Moqvist, & T. Popkewitz (Eds.), *Governing children, families & education: Restructuring the welfare state* (pp. 3–31). New York: Palgrave Macmillan.

Bowe, R., & Ball, S. (1992). *Reforming education & changing schools: Case studies in policy sociology.* London: Routledge.

Bruner, J. (1980). *Under five in Britain.* Ypsilanti, MI: High/Scope Press.

Curtis, P. (2003, August 2). Tony Blair on education [Electronic version]. *Guardian Unlimited.* Retrieved August 23, 2004, from http://education.guardian.co.uk/print/0%2C3858%2C4725045–110908%2C00.htmlDahlberg, G., Moss, P., & Pence, A. (Eds.). (1999). *Beyond quality in early childhood education and care: Postmodern perspectives.* London: Falmer Press.

Deleuze, G. (1988). *Foucault* (S. Hand, Trans.). Minneapolis, MN: University of Minnesota Press. (Original work published 1986)

Department for Education and Skills website. (n.d.). Retrieved February 27, 2004, from http://www.dfes.gov.U.K./educationact2002/

Education Act 2002 [c. 32] (2002). London: The Parliamentary Bookshop.

Foucault, M. (1973). *The birth of the clinic: An archeology of medical perception.* (A. M. Sheridan, Trans.). London: Tavistock Publications Limited. (Original work published 1963)

Foucault, M. (1977). *Discipline & punish: The birth of the prison* (2nd ed.). (A. Sheridan, Trans.). New York: Random House. (Original work published 1973)

Foucault, M. (1990). *The history of sexuality: Volume 1: An introduction* (R. Hurley, Trans.). New York: Vintage Books. (Original work published 1976)

Giddens, A. (2000). *The Third Way and its critics.* Cambridge: Polity Press.

Giddens, A. (2003). *The progressive manifesto.* Cambridge: Polity Press.

Gordon, C. (Ed.). (1980). *Power/knowledge: Selected interviews & other writings 1972–1977* (A. Fontana & P. Pasquino, Trans.) New York: Random House. (Original work published 1977)

Hacking, I. (1990). *The taming of chance.* Cambridge: Cambridge University Press.

Hatch, J. (2002). Accountability shovedown: Resisting the standards movement in early childhood education. *Phi Delta Kappan, 83* (6), 457–62.

Harbison, F., & Meyers, C. (1964). *Education, manpower, and economic growth; strategies of human resource development.* New York: McGraw-Hill.

Isaac, J. (2001). The road (not?) taken: Anthony Giddens, the Third Way, and the future of social democracy. *Dissent,* 100–108, 112. Kessler, S., & Swadener, B. (1992). *Reconceptualizing the early childhood curriculum: Beginning the dialogue.* New York: Teachers College Press.

Lightfoot, T. (2001). Education as literature: Tracing our metaphoric understandings of second language learners. Unpublished doctoral dissertation, University of Wisconsin-Madison.

Mahoney, P., & Hextall, I. (1997). *The policy context and impact of the TTA: A summary.* New York: RouledgeFalmer.

Martin, L., Gutman, H., & Hutton, P. (Eds.). (1988). *Technologies of the self.* Amherst, MA: University of Massachusetts Press.

Miliband, D. (2002). Speech presented at the 10 October 2002, Annual Meeting of QCA London. Retrieved October 22, 2005 from http://www.dfes.gov.U.K./speeches/search_detail.cfm?ID=41

Morris, E. (2002). The London challenge. Speech presented at the South Camden Community School on July 1. Retrieved October 22, 2005 from http://www.dfes.gov.U.K./speeches/search_detail.cfm?ID=38

Moss, P., & Penn, H. (Eds.). (1996). *Transforming nursery education*. London: Paul Chapman.

Office for Standards in Education /Ofsted (n.d.). *How we work*. Retrieved August 3, 2005 from http://www.ofsted.gov.U.K./howwework/

Osgood, J. (2004). Time to get down to business? The responses of early years practioners to entrepreneurial approaches to professionalism. *Journal of Early Childhood Research, 2* (1), 5024.

Paige, R. (2002). *Administration hails comprehensive education plan*. Address to Improving America's Schools Conference; Remarks of U. S. Secretary of Education Rod Paige on December 19, 2001. Retrieved September 23, 2005 from http://usinfo.state.gov/usa/edu/s121901.htm

Rose, N. (1999a). *Governing the soul: The shaping of the private self* (2nd ed.). London: Free Association Books.

Rose, N. (1999b). *Powers of freedom: Reframing political thought*. New York: Cambridge University Press.

Said, E. (1978). *Orientalism*. New York: Pantheon Books.

Salisbury, J., & Riddell, C. (Eds.). (2000). *Gender, policy & educational change: Shifting agendas in the U.K. and Europe*. New York: Routledge.

Swadener, B. & Lubeck, S. (Eds.). (1995). *Children and families "at promise": Deconstructing the discourse of risk*. New York: State University of New York Press.

Tabachnick, B. R., & Bloch, M. N. (1995). Learning in and out of School: Critical Perspectives on the Theory of Cultural Compatibility. In B. Swadener, and S. Lubeck (Eds.). *Children and Families "at Promise": the Social Construction of Risk* (pp. 187–209). New York: State University of New York Press.

United Kingdom Parliament. (1999, December 16). Report from the Minister for School Standards (Ms Estelle Morris). Members in the Commons Hansard Debates text for Thursday 16 Dec 1999. (Column: 111WH, Child Care [16 Dec 1999]). Retrieved on November 19, 2005, from http://www.publications.parliament.U.K./pa/cm199900/cmhansrd/vo991216/hallindx/91216-x.htm

Wegg-Prosser, B. (2002, June 11). Back in Focus [Electronic version]. *Guardian Unlimited*. Retrieved October 6, 2005, from http:// education.guardian.co.uk/print/0,3858,4430842–108228.00.html

Contributors

Susan Matoba Adler is Assistant Professor of Early Childhood Education and faculty of the Asian American Studies Program at the University of Illinois at Urbana-Champaign. Her research is on Asian American families, multicultural education, parent involvement of Asian Americans, and the racial-ethnic socialization of Asian American and Mixed-heritage children.

Marianne Bloch is Professor of Early Childhood and Elementary Education in the Department of Curriculum and Instruction, University of Wisconsin-Madison. She has done research on comparative and international education related to early education, child care, gender and education, and welfare state restructuring in relation to education and child care policy. Her theoretical interests focus on critical, feminist and poststructural theories of childhood, families, education and care.

Devorah Kennedy is currently a Dissertator at University of Wisconsin-Madison. Her interests include early childhood education, curriculum theory, postmodernism, postcolonialism, and Jewish studies.

I-Fang Lee is an Assistant professor at the Hong Kong Institute of Education. Her research interests include post-structural theories in curriculum and educational reforms issues in the field of early childhood education and care.

Theodora Lightfoot is an Assistant Professor of Education at the University of Illinois at Chicago. Her interests include the education of language minority students and postmodern and postcolonial theory.

Ana Laura Godinho Lima is an Assistant Professor of Psychology, Education and Contemporary Issues at the School of Arts, Sciences and Humanities at the University of São Paulo, Brazil. Her research interests include postmodernism, theories in history of education, psychology of education, and early childhood education.

Nancy Pauly is Assistant Professor in the Art Education Program at the University of New Mexico-Albuquerque, New Mexico. Her research interests include visual culture studies, multicultural art education, and social construction knowledge. She currently studies the ways young children construct meanings through television using multimodal literacies.

Ruth Peach is a Ph.D. candidate doctoral candidate in early childhood education policy in the Department of Curriculum and Instruction at the University of Wisconsin-Madison and an adjunct Faculty member at National-Lovis University. Her research interests include postmodern theories in early childhood education policy, educational philosophy and policy, and international discourses in education.

Karen S. Pena is a Ph.D. candidate in Educational Theory at the University of Wisconsin-Madison. Her research interests include postmodern thought in curriculum theory, philosophy, history of education, and cultural analysis.

Jie Qi is an Associate Professor of Education and Japanese at the Utsunomiya University, Japan. Her research interests include postmodernism theories in teacher education and multicultural education, educational philosophy and political sociology of education.

Dar Weyenberg is a doctoral candidate in the Department of Curriculum and Instruction at the University of Wisconsin-Madison. Her main research interests are histories of health education and governing practices.

Index